Leckie × Leckie

Scotland's leading educational publishers

KT-221-221

National 5
MATHS
SUCCESS GUIDE

N5 MATHS
SUCCESS GUIDE

Ken Nisbet

© 2013 Leckie & Leckie Ltd

001/31102013

10 9 8 7 6 5 4

All rights reserved. No part of this publication may be reproduced, stored in a retrieval system, or transmitted in any form or by any means, electronic, mechanical, photocopying, recording or otherwise, without the prior written permission of the Publisher or a licence permitting restricted copying in the United Kingdom issued by the Copyright Licensing Agency Ltd., 90 Tottenham Court Road, London W1T 4LP.

The author asserts his moral rights to be identified as the author of this work.

ISBN 9780007504671

Published by
Leckie & Leckie Ltd
An imprint of HarperCollins*Publishers*
Westerhill Road, Bishopbriggs, Glasgow, G64 2QT
T: 0844 576 8126   F: 0844 576 8131
leckieandleckie@harpercollins.co.uk   www.leckieandleckie.co.uk

Special thanks to
Ink Tank (cover design); QBS (layout and illustration); Jill Laidlaw (copy editing); Delphine Lawrance (proofreading); Roda Morrison (proofreading)

A CIP Catalogue record for this book is available from the British Library.

Acknowledgements
Whilst every effort has been made to trace the copyright holders, in cases where this has been unsuccessful, or if any have inadvertently been overlooked, the Publishers would gladly receive any information enabling them to rectify any error or omission at the first opportunity.

Printed in Italy by Grafica Veneta S.p.A.

Contents

Introduction

Contents

Unit 3 Applications

Assessment solutions   www.leckieandleckie.co.uk/n5mathssuccess

This book, your course and your exam

Using this book

TOP TIP

Always revisit questions you failed to solve, wait a few days, then try to solve them again. Use this guide to help!

About this book

This Success Guide was written to help you pass the National 5 Mathematics course. It will help you pass both the unit tests and the end-of-course exam. The book follows the unit structure of the course as detailed in the National Course Specification Document.

However, you don't become good at mathematics just by reading books – although they will give you the knowledge and skills you need to do so. The more you practise, the better you become – so use this book to start you off, then get out there and start problem-solving!

Top tips

Tips on mathematics and exam techniques can be found throughout the book: make sure you read and learn them.

Quick tests

These are at the end of each topic. If you have difficulty with these questions after revising the topic, then you should practise the topic more. Answers are provided at the back of the book.

Sample assessment questions

TOP TIP

Use the Quick tests and sample assessment questions to identify your strengths and weaknesses.

At the end of each unit there are typical unit and end-of-course assessment questions. Detailed solutions to these questions are provided at www.leckieandleckie.co.uk/n5mathssuccess. You should spend time attempting these questions and then compare your solutions carefully with the given solutions.

Exam practice

TOP TIP

When revising maths **all** your time should be spent doing questions. Use notes to help you understand how to solve questions.

The best practice for tests and exams is to sit a practice paper under exam conditions at home and then compare your solutions with exemplar solutions. You should obtain Leckie and Leckie National 5 Mathematics Practice Papers for this purpose. They contain typical exam papers with detailed solutions, helpful comments and useful guidance. See www.leckieandleckie.co.uk. The last few years', actual exams can be downloaded directly from the SQA website at www.sqa.org.uk/pastpapers.

Getting more help

You can always go back to your textbook, your notes or your teacher for more examples and explanations if there's anything you're not sure about. If you are learning on your own, you may need to find a knowledgeable friend to help you out occasionally.

The course structure

The National 5 Mathematics course consists of three units:

Mathematics: Expressions and Formulae	Unit 1 of this guide
Mathematics: Relationships	Unit 2 of this guide
Mathematics: Applications	Unit 3 of this guide

Success at this course will allow you to progress to the Higher Mathematics course.

The assessment structure

To gain a course award you have to pass all three unit tests as well as the end-of-course exam. You should realise that the unit tests contain only the easier bits of the course and that unit test questions are very predictable. So if you get very good at unit test questions this does not necessarily mean you will pass the end-of-course exam.

Your unit tests will be sat on a unit-by-unit basis when your teacher decides you are ready. There are no grades in these, only a pass or fail. You will be allowed to use a calculator during all your unit tests. Normally only one attempt at a resit is allowed should you fail a unit test.

The end-of-course exam will consist of two papers as follows:

Paper 1 (non-calculator)	**60 minutes**	**worth 40 marks**
Paper 2 (calculator allowed)	**90 minutes**	**worth 50 marks**

Depending how well you did in this end-of-course exam (and providing you passed all three unit tests) you will be awarded a grade A-D. Specimen question papers can be downloaded directly from the SQA website at www.sqa.org.uk.

The formulae list

The following formulae list will be available to you during your assessments:

The roots of $ax^2 + bx + c = 0$ are $x = \dfrac{-b \pm \sqrt{(b^2 - 4ac)}}{2a}$

Sine rule: $\dfrac{a}{\sin A} = \dfrac{b}{\sin B} = \dfrac{c}{\sin C}$

Cosine rule: $a^2 = b^2 + c^2 - 2bc \cos A$ or $\cos A = \dfrac{b^2 + c^2 - a^2}{2bc}$

Area of a triangle: $\text{Area} = \frac{1}{2}ab \sin C$

Volume of a sphere: $\qquad V = \frac{4}{3}\pi r^3$

Volume of a cone: $\qquad V = \frac{1}{3}\pi r^2 h$

Volume of a pyramid: $\qquad V = \frac{1}{3}Ah$

Standard deviation: $\qquad s = \sqrt{\dfrac{\sum(x-\bar{x})^2}{n-1}} = \sqrt{\dfrac{\sum x^2 - (\sum x)^2/n}{n-1}}$, where n is the sample size.

Working with surds

TOP TIP

All integers are rational numbers, e.g.

$$6 = \frac{6}{1} \quad -2 = \frac{-2}{1}$$

Types of numbers

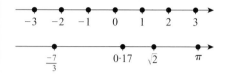

You are familiar with the integers. Here a few of them are shown on a number line. There are many more numbers than just the integers. Every point on this number line corresponds to a real number.

There are two types of real numbers:

Rational numbers

$$\frac{1}{2} \quad -\frac{7}{3} \quad \frac{6}{1} \quad \frac{143}{23} \quad -\frac{8}{16}$$

These are the 'ratio' of two integers: $\frac{m}{n}$ where $n \neq 0$.

Irrational numbers

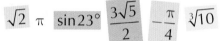

$$\sqrt{2} \quad \pi \quad \sin 23° \quad \frac{3\sqrt{5}}{2} \quad -\frac{\pi}{4} \quad \sqrt[3]{10}$$

These cannot be written as the 'ratio' of two integers.

What is a square root?

The square root of n is a number, written $\sqrt{n}$, that gives n when squared.

Basic result: $\sqrt{n}^2 = \sqrt{n} \times \sqrt{n} = n$

Examples

$\sqrt{9} = 3$ since $3^2 = 3 \times 3 = 9$

$\sqrt{25} = 5$ since $5^2 = 5 \times 5 = 25$

What is a surd?

TOP TIP

The root of a SQUARE NUMBER is not a surd.

A surd is an irrational root of a rational number – for our purposes a square root of a positive integer.

Examples of surds

$$\sqrt{2} \quad \sqrt{3} \quad \sqrt{6} \quad \sqrt{5} \quad \sqrt{7}$$

$\sqrt{4}$ is not a surd since it is not irrational – its value is 2.

Calculator tip

Irrational numbers like π, $\sqrt{2}$ and $\sin 14°$ have decimal values that never end and never repeat. You never know their values exactly.

You can never know the exact value of a surd.

TOP TIP

$\sqrt{4} + \sqrt{9} = 2 + 3 = 5$

$\sqrt{4+9} = \sqrt{13} \neq 5$

How do surds behave?

Compare:

$\sqrt{2+3}$ and $\sqrt{2}+\sqrt{3}$

A quick calculator check gives:

$\sqrt{2+3} = \sqrt{5} = 2 \cdot 23...$

Not the same

$\sqrt{2} + \sqrt{3} = 1 \cdot 41... + 1 \cdot 73... = 3 \cdot 14...$

$\sqrt{2 \times 3} = \sqrt{6} = 2 \cdot 44...$

$\sqrt{2 \times 3}$ and $\sqrt{2} \times \sqrt{3}$

$\sqrt{2} \times \sqrt{3} = 1 \cdot 41... \times 1 \cdot 73... = 2 \cdot 44...$

They might be the same!

$\sqrt{\dfrac{2}{3}}$ and $\dfrac{\sqrt{2}}{\sqrt{3}}$

$\sqrt{\dfrac{2}{3}} = 0 \cdot 81...$

$\dfrac{\sqrt{2}}{\sqrt{3}} = \dfrac{1 \cdot 41...}{1 \cdot 73...} = 0 \cdot 81...$

They might be the same!

The general rules that are true are:

$\sqrt{a \times b} = \sqrt{a} \times \sqrt{b}$ and $\sqrt{\dfrac{a}{b}} = \dfrac{\sqrt{a}}{\sqrt{b}}$ These can be proved.

Note: $\sqrt{a}^2 = \sqrt{a} \times \sqrt{a} = \sqrt{a \times a} = \sqrt{a^2} = a$

TOP TIP

When simplifying a surd look for the highest square number factor.

Simplifying surds

To simplify $\sqrt{108}$ find a factor of 108 that is a square number, e.g. 9. You know $108 = 9 \times 12$.

So $\sqrt{108} = \sqrt{9 \times 12} = \sqrt{9} \times \sqrt{12} = 3 \times \sqrt{12}$.

The square numbers

1	16	36	81
4	9	49	100
	25	64	

Just as $3 \times x$ is written $3x$ so $3 \times \sqrt{12}$ is written $3\sqrt{12}$.
Now 12 has a square factor of 4.
So $3\sqrt{12} = 3 \times \sqrt{4 \times 3} = 3 \times \sqrt{4} \times \sqrt{3}$

$= 3 \times 2 \times \sqrt{3} = 6 \times \sqrt{3} = 6\sqrt{3}$

Here is a quicker method:

$\sqrt{108} = \sqrt{36 \times 3} = \sqrt{36} \times \sqrt{3} = 6 \times \sqrt{3} = 6\sqrt{3}$

When the number under the root sign is as small as possible (no more square factors other than 1) you have fully simplified the surd.

Examples

Simplify: (a) $\sqrt{96}$ (b) $\sqrt{\dfrac{81}{100}}$

(a) $\sqrt{96} = \sqrt{16 \times 6} = \sqrt{16} \times \sqrt{6} = 4 \times \sqrt{6} = 4\sqrt{6}$

(b) $\sqrt{\dfrac{81}{100}} = \dfrac{\sqrt{81}}{\sqrt{100}} = \dfrac{9}{10} = 0 \cdot 9$

Rationalising the denominator

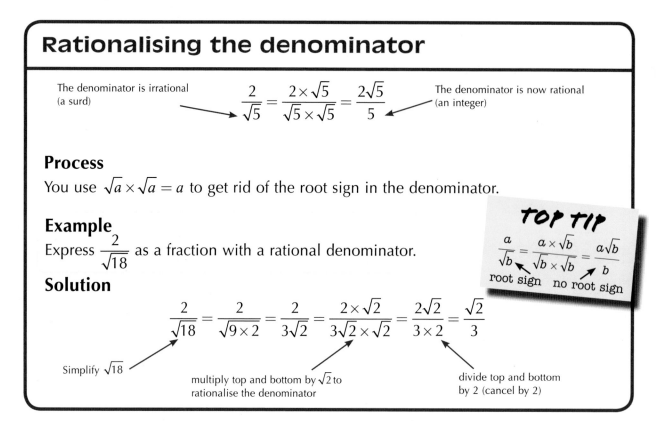

The denominator is irrational (a surd)

$$\frac{2}{\sqrt{5}} = \frac{2 \times \sqrt{5}}{\sqrt{5} \times \sqrt{5}} = \frac{2\sqrt{5}}{5}$$

The denominator is now rational (an integer)

Process

You use $\sqrt{a} \times \sqrt{a} = a$ to get rid of the root sign in the denominator.

Example

Express $\dfrac{2}{\sqrt{18}}$ as a fraction with a rational denominator.

TOP TIP

$$\frac{a}{\sqrt{b}} = \frac{a \times \sqrt{b}}{\sqrt{b} \times \sqrt{b}} = \frac{a\sqrt{b}}{b}$$

root sign   no root sign

Solution

$$\frac{2}{\sqrt{18}} = \frac{2}{\sqrt{9 \times 2}} = \frac{2}{3\sqrt{2}} = \frac{2 \times \sqrt{2}}{3\sqrt{2} \times \sqrt{2}} = \frac{2\sqrt{2}}{3 \times 2} = \frac{\sqrt{2}}{3}$$

Simplify $\sqrt{18}$

multiply top and bottom by $\sqrt{2}$ to rationalise the denominator

divide top and bottom by 2 (cancel by 2)

Estimation

Estimating values without a calculator is an important skill. Before any calculation you should make a rough estimate.

$$\sqrt{1} = 1$$

$$\left.\begin{array}{l}\sqrt{2} = 1 \cdot 41... \\ \sqrt{3} = 1 \cdot 73...\end{array}\right\} \begin{array}{l}\text{decimal value} \\ \text{starts } 1 \cdot ...\end{array}$$

$$\sqrt{4} = 2$$

$$\left.\begin{array}{l}\sqrt{5} = 2 \cdot 24... \\ \sqrt{6} = 2 \cdot 45... \\ \sqrt{7} = 2 \cdot 65... \\ \sqrt{8} = 2 \cdot 83...\end{array}\right\} \begin{array}{l}\text{decimal value} \\ \text{starts } 2 \cdot ...\end{array}$$

$$\sqrt{9} = 3$$

Explain why $\sqrt{53}$ starts $7 \cdot ...$
(think of 49 and 64)

Example

Estimate the value of $\dfrac{8}{\sqrt{5}}$.

Solution

First rationalise the denominator:

$$\frac{8}{\sqrt{5}} = \frac{8 \times \sqrt{5}}{\sqrt{5} \times \sqrt{5}} = \frac{8\sqrt{5}}{5}$$

Now $\sqrt{5}$ lies between 2 ($\sqrt{4}$) and 3 ($\sqrt{9}$)
(closer to 2)

So $8\sqrt{5}$ lies between 16 and 24
(closer to 16)

So $\dfrac{8\sqrt{5}}{5}$ lies between 3 and 5
(closer to 3)

Rough estimate is $3 \cdot 5$ say

(Actual value is $3 \cdot 57...$)

Further simplification

If you think of $\sqrt{2}$ as an unknown number like x then you can compare

$$2\sqrt{2} + 3\sqrt{2} = 5\sqrt{2}$$

with $2x + 3x = 5x$

Similarly by comparing with $x + 3y + 4x - y$ you can simplify

$$\sqrt{3} + 3\sqrt{2} + 4\sqrt{3} - \sqrt{2}$$
$$= \sqrt{3} + 4\sqrt{3} + 3\sqrt{2} - \sqrt{2}$$
$$= 5\sqrt{3} + 2\sqrt{2}$$

TOP TIP

$\sqrt{2} + \sqrt{3}$ cannot be simplified (compare $x + y$).

Examples
Simplify:

(a) $2\sqrt{5} - 3\sqrt{2} + \sqrt{5} + 4\sqrt{2}$ (b) $\sqrt{2} + \dfrac{2}{\sqrt{2}}$

Solutions
(a) Rewrite as

$$2\sqrt{5} + \sqrt{5} + 4\sqrt{2} - 3\sqrt{2} = 3\sqrt{5} + \sqrt{2}$$

(compare $2x - 3y + x + 4y$)

(b) $\sqrt{2} + \dfrac{2}{\sqrt{2}} = \sqrt{2} + \dfrac{2 \times \sqrt{2}}{\sqrt{2} \times \sqrt{2}} = \sqrt{2} + \dfrac{2\sqrt{2}}{2} = \sqrt{2} + \sqrt{2} = 2\sqrt{2}$

Quick Test 1

1. Simplify:

    a) $\sqrt{45}$ b) $\sqrt{216}$ c) $\sqrt{\dfrac{49}{64}}$

2. Express as a fraction with a rational denominator and simplify if possible:

    a) $\dfrac{1}{\sqrt{3}}$ b) $\dfrac{10}{\sqrt{5}}$ c) $\dfrac{14}{\sqrt{98}}$

3. Simplify:

    a) $4\sqrt{7} - \sqrt{2} - 3\sqrt{7} + \sqrt{2}$ b) $3\sqrt{5} - \dfrac{10}{\sqrt{5}}$

Working with decimals and integers (a review)

Decimal places

Decimal places are counted immediately to the right of the decimal point:

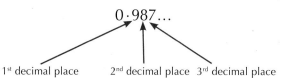

$$0 \cdot 987\ldots$$

1st decimal place   2nd decimal place   3rd decimal place

Examples
0·02 kg is written to 2 decimal places.
15·6230 m is written to 4 decimal places.

Significant figures

For the measurements

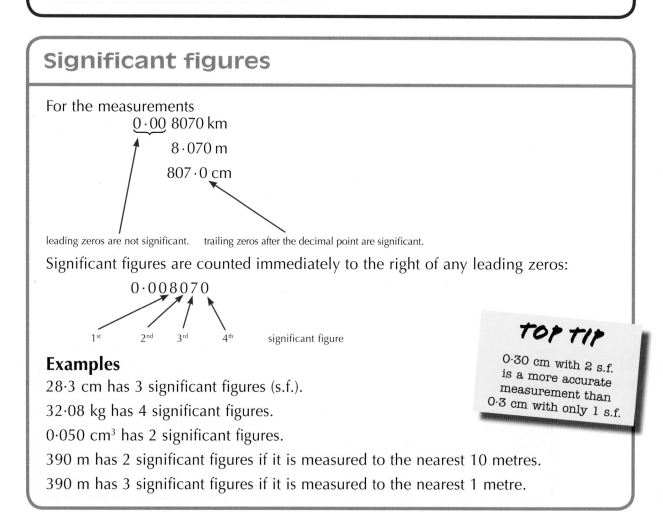

$$\underline{0 \cdot 00}\, 8070 \text{ km}$$
$$8 \cdot 070 \text{ m}$$
$$807 \cdot 0 \text{ cm}$$

leading zeros are not significant.   trailing zeros after the decimal point are significant.

Significant figures are counted immediately to the right of any leading zeros:

$$0 \cdot 008070$$

1st   2nd   3rd   4th   significant figure

TOP TIP

0·30 cm with 2 s.f. is a more accurate measurement than 0·3 cm with only 1 s.f.

Examples
28·3 cm has 3 significant figures (s.f.).

32·08 kg has 4 significant figures.

0·050 cm³ has 2 significant figures.

390 m has 2 significant figures if it is measured to the nearest 10 metres.

390 m has 3 significant figures if it is measured to the nearest 1 metre.

Rounding measurements

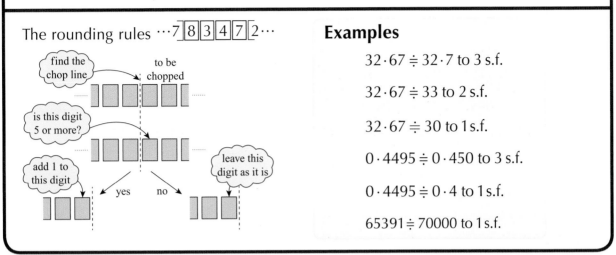

The rounding rules ···7⬚8⬚3⬚4⬚7⬚2···   **Examples**

find the chop line — to be chopped

is this digit 5 or more?

add 1 to this digit   yes   no   leave this digit as it is

$32 \cdot 67 \doteqdot 32 \cdot 7$ to 3 s.f.

$32 \cdot 67 \doteqdot 33$ to 2 s.f.

$32 \cdot 67 \doteqdot 30$ to 1 s.f.

$0 \cdot 4495 \doteqdot 0 \cdot 450$ to 3 s.f.

$0 \cdot 4495 \doteqdot 0 \cdot 4$ to 1 s.f.

$65391 \doteqdot 70000$ to 1 s.f.

Advice on rounding

- It is useful to draw number lines.
- Never round values in a calculation until the final answer is obtained. Calculating with rounded values leads to inaccurate results.
- Don't leave more significant figures in an answer than there were in the measurements you used for the calculation!

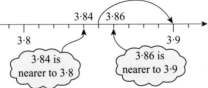

3·84   3·86

3·8   3·9

3·84 is nearer to 3·8

3·86 is nearer to 3·9

Note: The midpoint number 3·85 is rounded up to 3·9.

Adding and subtracting integers

Use the number line:

$-2 - 3 = -5$ $-2 + 3 = 1$
$-2 + (-3) = -5$ $-2 - (-3) = 1$

-5   -4   -3   -2   -1   0   1

Notice that:

Adding a negative is the same as subtracting so $+ (-3)$ gives -3.

Subtracting a negative is the same as adding so $- (-3)$ gives $+3$.

Example

Calculate the next three terms in the sequence 5, −4, 1, ... where the rule is 'add the two previous terms to get the next term'.

Solution

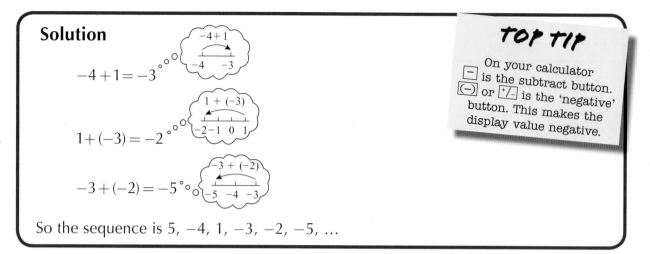

$-4 + 1 = -3$

$1 + (-3) = -2$

$-3 + (-2) = -5$

TOP TIP

On your calculator ⊟ is the subtract button. ⊝ or +/– is the 'negative' button. This makes the display value negative.

So the sequence is 5, −4, 1, −3, −2, −5, …

Multiplying and dividing integers

Use the rules:

$\left.\begin{array}{l} \text{positive} \times \text{positive} \\ \text{negative} \times \text{negative} \end{array}\right\}$ same signs give positive

$\left.\begin{array}{l} \text{negative} \times \text{positive} \\ \text{positive} \times \text{negative} \end{array}\right\}$ different signs give negative

Exactly the same rules apply to dividing two integers.

Examples
Calculate:

1. $(-3)^2$ 2. $(-2)^4$

3. $\dfrac{-3 + (-5) \times 3}{-2 \times 3}$

Solutions

1. $(-3)^2 = -3 \times (-3) = 9$

2. $(-2)^4 = -2 \times (-2) \times (-2) \times (-2)$

   $= -2 \times (-2) \times 4$ (since neg × neg is pos)

   $= -2 \times (-8)$ (since neg × pos is neg)

   $= \mathbf{16}$ (again since neg × neg is pos)

3. $\dfrac{-3 + (-5) \times 3}{-2 \times 3} = \dfrac{-3 + (-15)}{-6} = \dfrac{-18}{-6} = \mathbf{3}$

   (neg ÷ neg is pos)

Quick Test 2

1. Round:

    a) 23·55 m to 1 s.f.   b) 0·0950 km to 2 s.f.

2. Calculate:

    a) $-3 \times 2 \times (-4)$ b) $-2 - (-5)$ c) $(-1)^2 + (-2)^2$

3. Calculate:

    a) $\dfrac{6}{-2}$ b) $\dfrac{2 \times (-3)}{-1}$ c) $\dfrac{-2 + (-3) + 1}{-1 \times (-2)}$

Working with fractions (a review)

TOP TIP

You must be able to work with fractions without a calculator.

Simplifying fractions

$\frac{1}{2} = \frac{2}{4} = \frac{3}{6} = \frac{4}{8} = \ldots$ all these fractions are equivalent.

You can multiply or divide the top (numerator) and the bottom (denominator) of a fraction by the same number to get an equivalent fraction.

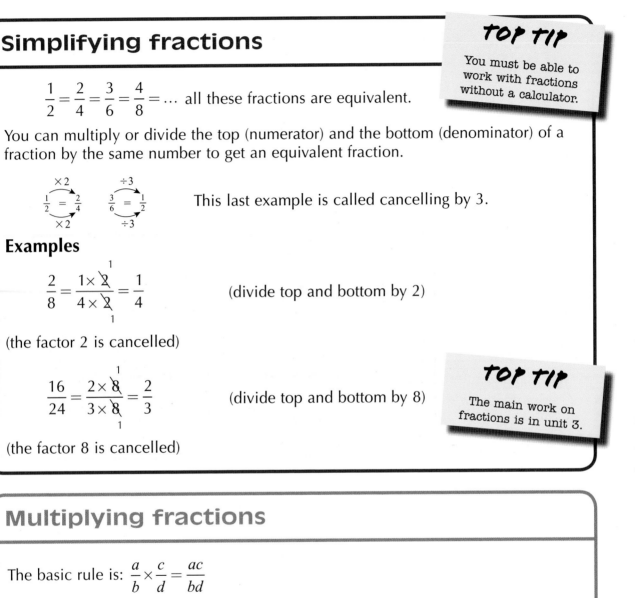

This last example is called cancelling by 3.

Examples

$$\frac{2}{8} = \frac{1 \times \cancel{2}^{1}}{4 \times \cancel{2}_{1}} = \frac{1}{4}$$ (divide top and bottom by 2)

(the factor 2 is cancelled)

$$\frac{16}{24} = \frac{2 \times \cancel{8}^{1}}{3 \times \cancel{8}_{1}} = \frac{2}{3}$$ (divide top and bottom by 8)

TOP TIP

The main work on fractions is in unit 3.

(the factor 8 is cancelled)

Multiplying fractions

The basic rule is: $\frac{a}{b} \times \frac{c}{d} = \frac{ac}{bd}$

Multiply the two numerators and multiply the two denominators.

Examples

$$\frac{2}{3} \times \frac{7}{9} = \frac{2 \times 7}{3 \times 9} = \frac{14}{27}$$

$$\frac{3}{7} \times \frac{14}{15} = \frac{3 \times 14}{7 \times 15} = \frac{\cancel{3} \times 2 \times \cancel{7}}{\cancel{7} \times \cancel{3} \times 5} = \frac{2}{5}$$

(the factors 3 and 7 can be cancelled)

Dividing fractions

The method explained here will be useful when you work with algebraic fractions.

$$2 \div \frac{1}{3} = \frac{2}{\frac{1}{3}} = \frac{2 \times 3}{\frac{1}{3} \times 3} = \frac{6}{1} = 6$$ Multiply top and bottom by 3 to get rid of $\frac{1}{3}$.

$$\frac{2}{7} \div 5 = \frac{\frac{2}{7}}{5} = \frac{\frac{2}{7} \times 7}{5 \times 7} = \frac{2}{35}$$ Multiply top and bottom by 7 to get rid of $\frac{2}{7}$

since $\frac{2}{7} \times 7 = \frac{2}{7} \times \frac{7}{1} = \frac{2 \times 7}{7 \times 1} = \frac{2}{1} = 2$

$$\frac{1}{4} \div \frac{2}{5} = \frac{\frac{1}{4}}{\frac{2}{5}} = \frac{\frac{1}{4} \times 4 \times 5}{\frac{2}{5} \times 4 \times 5} = \frac{5}{8}$$ Multiply top and bottom by 4 to get rid of $\frac{1}{4}$ and by 5 to get rid of $\frac{2}{5}$.

(note: $\frac{2}{5} \times \frac{4}{1} \times \frac{5}{1} = \frac{2 \times 4 \times 5}{5 \times 1 \times 1} = 2 \times 4 = 8$)

Adding and subtracting fractions

Your aim is to get the denominators the same, e.g. $\frac{2}{3} + \frac{1}{5}$ (2 thirds and 1 fifth).

This changes to $\frac{2 \times 5}{3 \times 5} + \frac{1 \times 3}{5 \times 3} = \frac{10}{15} + \frac{3}{15}$ (10 fifteenths and 3 fifteenths)

$$= \frac{10 + 3}{15} = \frac{13}{15}$$ (adding to get 13 fifteenths)

Similarly $\frac{5}{6} - \frac{3}{4} = \frac{5 \times 2}{6 \times 2} - \frac{3 \times 3}{4 \times 3} = \frac{10}{12} - \frac{9}{12} = \frac{1}{12}$ (10 twelfths minus 9 twelfths giving 1 twelfth).

Working with indices

What is an index?

$2^3 = 2 \times 2 \times 2 = 8$ 3 is the index. If the index n is a positive integer then it counts the repeated multiplication factor:

$$a^n = \underbrace{a \times a \times .... \times a}_{n \text{ factors}}$$

TOP TIP

Sometimes 'power' is used instead of 'index'.

However indices need not just be positive integers as this pattern suggests:

$$\cdots \; 2^3 \quad 2^2 \quad 2^1 \quad 2^0 \quad 2^{-1} \quad 2^{-2} \quad 2^{-3} \; \cdots$$

TOP TIP

Another name for an index is an exponent.

The rules of indices

Rule	Comments	Examples
$x^m \times x^n = x^{m+n}$	When multiplying, the indices are added. Note: not in the case $x^m \times y^n$!	$a^2 \times a^3 = a^{2+3} = a^5$
$\dfrac{x^m}{x^n} = x^{m-n}$	When dividing, the indices are subtracted.	$\dfrac{c^7}{c^3} = c^{7-3} = c^4$
$(x^m)^n = x^{mn}$	When raising a power to a power, multiply the indices.	$(y^3)^4 = y^{3 \times 4} = y^{12}$
$x^0 = 1$	Any number or expression (other than zero) raised to the power zero gives 1.	$2^0 = 1 \qquad \left(\dfrac{1}{2}\right)^0 = 1$ $\quad (a+b)^0 = 1$
$x^{-n} = \dfrac{1}{x^n}$	Something to a negative power can be rewritten as 1 divided by the same thing to the positive power.	$a^{-1} = \dfrac{1}{a^1} = \dfrac{1}{a} \quad a^{-3} = \dfrac{1}{a^3}$

Fractional indices

TOP TIP

$a^{\frac{power}{root}}$

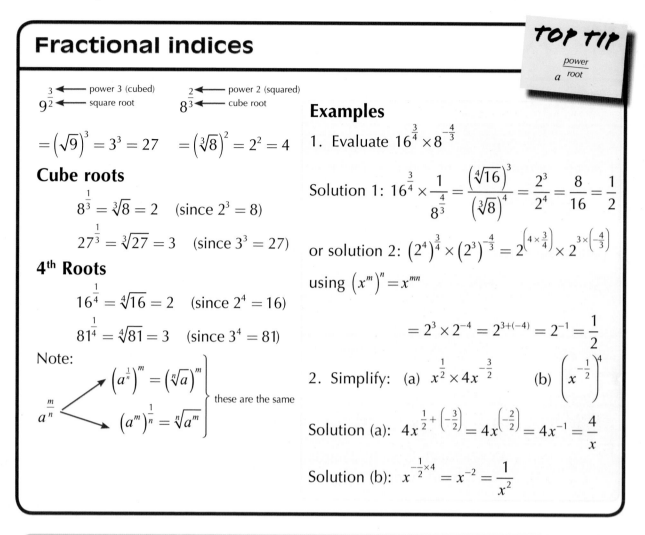

$9^{\frac{3}{2}}$ power 3 (cubed)   square root

$8^{\frac{2}{3}}$ power 2 (squared)   cube root

$= \left(\sqrt{9}\right)^3 = 3^3 = 27$ $= \left(\sqrt[3]{8}\right)^2 = 2^2 = 4$

Cube roots

$$8^{\frac{1}{3}} = \sqrt[3]{8} = 2 \quad (\text{since } 2^3 = 8)$$

$$27^{\frac{1}{3}} = \sqrt[3]{27} = 3 \quad (\text{since } 3^3 = 27)$$

4th Roots

$$16^{\frac{1}{4}} = \sqrt[4]{16} = 2 \quad (\text{since } 2^4 = 16)$$

$$81^{\frac{1}{4}} = \sqrt[4]{81} = 3 \quad (\text{since } 3^4 = 81)$$

Note:

$$a^{\frac{m}{n}} \left\{ \begin{array}{l} \left(a^{\frac{1}{n}}\right)^m = \left(\sqrt[n]{a}\right)^m \\ \left(a^m\right)^{\frac{1}{n}} = \sqrt[n]{a^m} \end{array} \right\} \text{ these are the same}$$

Examples

1. Evaluate $16^{\frac{3}{4}} \times 8^{-\frac{4}{3}}$

Solution 1: $16^{\frac{3}{4}} \times \dfrac{1}{8^{\frac{4}{3}}} = \dfrac{\left(\sqrt[4]{16}\right)^3}{\left(\sqrt[3]{8}\right)^4} = \dfrac{2^3}{2^4} = \dfrac{8}{16} = \dfrac{1}{2}$

or solution 2: $\left(2^4\right)^{\frac{3}{4}} \times \left(2^3\right)^{-\frac{4}{3}} = 2^{\left(4 \times \frac{3}{4}\right)} \times 2^{3 \times \left(-\frac{4}{3}\right)}$

using $\left(x^m\right)^n = x^{mn}$

$$= 2^3 \times 2^{-4} = 2^{3+(-4)} = 2^{-1} = \dfrac{1}{2}$$

2. Simplify: (a) $x^{\frac{1}{2}} \times 4x^{-\frac{3}{2}}$ (b) $\left(x^{-\frac{1}{2}}\right)^4$

Solution (a): $4x^{\frac{1}{2} + \left(-\frac{3}{2}\right)} = 4x^{\left(-\frac{2}{2}\right)} = 4x^{-1} = \dfrac{4}{x}$

Solution (b): $x^{-\frac{1}{2} \times 4} = x^{-2} = \dfrac{1}{x^2}$

What is scientific notation?

A number written as $a \times 10^n$ where $1 \le a < 10$ and n is an integer is in scientific notation or standard form, e.g. $2{\cdot}6 \times 10^3$ (2600).

Note: The value a (e.g. $2{\cdot}6$) must lie between 1 or 10 (or be equal to 1).

Example

Write $3{\cdot}28 \times 10^{-4}$ in normal decimal form.

Solution: $0{\cdot}000328$

Note: index -4 indicates to move the decimal point 4 places to the left.

Calculating with scientific notation

You need to be able to do simple calculations in scientific notation without the help of a calculator.

Note: $10^m \times 10^n = 10^{m+n}$ and $\dfrac{10^m}{10^n} = 10^{m-n}$

Examples
Calculate:

(a) $(3\cdot2 \times 10^{-3}) \times (4 \times 10^4)$ (b) $\dfrac{6 \times 10^2}{2 \times 10^{-3}}$ giving your answers in scientific notation.

Solutions

(a) $3\cdot2 \times 10^{-3} \times 4 \times 10^4 = 3\cdot2 \times 4 \times 10^{-3} \times 10^4 = 12\cdot8 \times 10^{-3+4} = 12\cdot8 \times 10^1$
$= 128 = 1\cdot28 \times 10^2$

(notice that $12\cdot8$ does not lie between 1 and 10 but $1\cdot28$ does).

(b) $\dfrac{6 \times 10^2}{2 \times 10^{-3}} = \dfrac{6}{2} \times \dfrac{10^2}{10^{-3}} = 3 \times 10^{2-(-3)} = 3 \times 10^{2+3} = 3 \times 10^5$

Note: A number in scientific notation can be entered into your calculator using the $\boxed{\times 10^x}$ (or $\boxed{\text{EE}}$ or $\boxed{\text{EXP}}$) key.

For example to enter $2\cdot6 \times 10^7$ you press:

$\boxed{2}\boxed{\cdot}\boxed{6}\boxed{\times 10^x}\boxed{7}$ (see your calculator manual for more details).

TOP TIP

$2\cdot38 \times 10^{-5}$ is written with 3 significant figures.

Quick Test (non-calculator) 3

1. Evaluate:

   a) $25^{\frac{3}{2}}$ b) $27^{-\frac{2}{3}}$ c) $9^{-\frac{1}{2}} \times 8^{\frac{2}{3}}$

2. Simplify:

   a) $\dfrac{x^2 x^4}{x^3}$ b) $\left(n^{-\frac{1}{3}}\right)^6$ c) $\dfrac{a^{\frac{1}{2}}}{a^{-\frac{3}{2}}}$

3. The universe is estimated to contain $1\cdot8 \times 10^{11}$ galaxies each containing 8×10^{11} stars. How many stars is this in total?

   Give your answer in scientific notation to 2 significant figures.

Working with algebraic terms (a review)

What are algebraic terms and expressions?

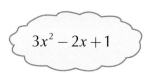

$3x^2 - 2x + 1$

This is an algebraic expression with three terms:

the x^2 term is $3x^2$. 3 is the coefficient of x^2.

the x term is $-2x$. -2 is the coefficient of x.

1 is called a constant term.

The letter x is called a variable and represents any number.

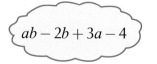

$ab - 2b + 3a - 4$

This is a four-term expression with two different variables, a and b.

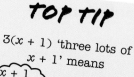

TOP TIP

$3(x + 1)$ 'three lots of $x + 1$' means

$x + 1$
$x + 1$
$x + 1$

giving $3x + 3$

Useful ways to think about terms

The term $3n$ means $3 \times n$ and it's useful to think: '3 lots of n'.

Multiplication	How to write	How to think
$3 \times n$	$3n$	'3 lots of n'
$2 \times n$	$2n$	'2 lots of n'
$1 \times n$	n	'1 lot of n'
$0 \times n$	0	'0 lots of n'
$-1 \times n$	$-n$	'minus 1 lot of n'
$-2 \times n$	$-2n$	'minus 2 lots of n'
$-3 \times n$	$-3n$	'minus 3 lots of n'

Other examples

a^3 means $a \times a \times a$ 'a cubed'

a^2 means $a \times a$ 'a squared'

$4a^2$ means $4 \times a \times a$ '4 lots of a squared'

$3(m + 1)$ '3 lots of $m + 1$'

TOP TIP

Multiplying together an **odd** number of negative terms gives a negative result.

Multiplying terms

You apply the same rules that you use for multiplying integers:

positive $\times$ positive
negative $\times$ negative
} same signs give positive

negative $\times$ positive
positive $\times$ negative
} different signs give negative

Examples

$-3a \times 4a = -12a^2$ (neg $\times$ pos)

$-2m \times (-n) = 2mn$ (neg $\times$ neg)

$-k \times (-k) \times (-3) = -3k^2$

$(-x)^2 = -x \times (-x) = x^2$

Adding and subtracting terms

Just as you would use a number line to add and subtract integers:

$$1 - 3 = -4 \qquad\qquad -1 + 3 = 2$$
$$1 + (-3) = -4 \qquad -1 - (-3) = 2$$

In the same way you use a number line to add and subtract algebraic terms:

$$-x - 3x = -4x \qquad\qquad -x + 3x = 2x$$
$$-x + (-3x) = -4x \qquad -x - (-3x) = 2x$$

Examples

1. $m - (-4m) = m + 4m = 5m$
   (subtracting a negative is the same as adding)

2. $-x - 3x = -4x$
   (compare this with:

   $$-1 - 3 = -4$$

)

3. $2n^2 + n - 5n^2 = -3n^2 + n$

   (2 lots of n^2 minus 5 lots of n^2, but note that an n^2 term and an n term cannot be added together)

TOP TIP

Only 'like' terms can be added or subtracted so $3a - 2a = a$ but $3a - 2b$ cannot be simplified.

Evaluating expressions

If the value of a variable is known then you can find the value of an expression in that variable.

For example if $x = 3$ then $8x^2 - 2x$ becomes $8 \times 3^2 - 2 \times 3 = 66$.

The expression $8x^2 - 2x$ has a value 66. It has been evaluated!

Example

Evaluate $2ab - a^2$ when $a = 7$ and $b = 10$.

Solution

$2ab - a^2 = 2 \times 7 \times 10 - 7^2 = 140 - 49 = 91$

Quick Test 4

1. a) What is the constant term in $3ab - 3a + 2b + 4$?

   b) What is the coefficient of x^2 in $3x^3 - x^2 + x$?

2. Simplify:

   a) $x \times (-2)$ b) $-n \times 3 \times (-n)$

   c) $3a - 6a$ d) $4b - 3a + b - 7a$

3. Evaluate:

   a) $5mn - n^2$ when $m = 6$ and $n = 3$

   b) $x^2 - 3x - 2$ when $x = -1$

Expanding brackets in algebraic expressions

Removing one pair of brackets

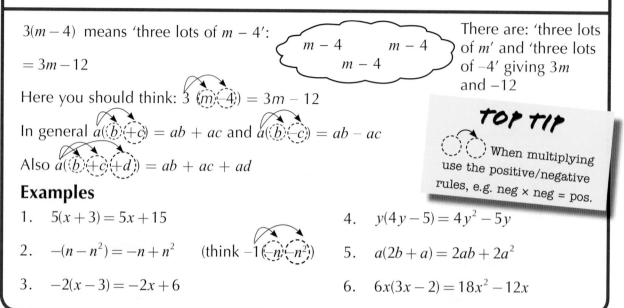

$3(m - 4)$ means 'three lots of $m - 4$':

$= 3m - 12$

Here you should think: $3(m - 4) = 3m - 12$

In general $a(b + c) = ab + ac$ and $a(b - c) = ab - ac$

Also $a(b + c + d) = ab + ac + ad$

There are: 'three lots of m' and 'three lots of -4' giving $3m$ and -12

TOP TIP

When multiplying use the positive/negative rules, e.g. neg × neg = pos.

Examples

1. $5(x + 3) = 5x + 15$

2. $-(n - n^2) = -n + n^2$ (think $-1(-n - n^2)$)

3. $-2(x - 3) = -2x + 6$

4. $y(4y - 5) = 4y^2 - 5y$

5. $a(2b + a) = 2ab + 2a^2$

6. $6x(3x - 2) = 18x^2 - 12x$

Removing two pairs of brackets

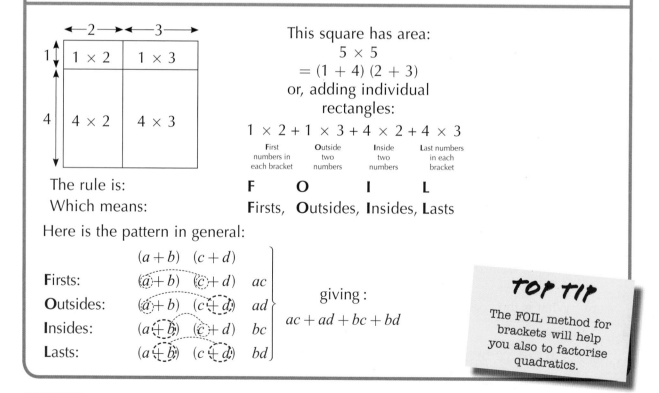

This square has area:
5×5
$= (1 + 4)(2 + 3)$
or, adding individual rectangles:
$1 \times 2 + 1 \times 3 + 4 \times 2 + 4 \times 3$

First numbers in each bracket | Outside two numbers | Inside two numbers | Last numbers in each bracket

The rule is: **F   O   I   L**

Which means: **F**irsts, **O**utsides, **I**nsides, **L**asts

Here is the pattern in general:

$(a + b)\ (c + d)$

Firsts: $(a + b)\ (c + d)$ ac

Outsides: $(a + b)\ (c + d)$ ad

Insides: $(a + b)\ (c + d)$ bc

Lasts: $(a + b)\ (c + d)$ bd

giving:
$ac + ad + bc + bd$

TOP TIP

The FOIL method for brackets will help you also to factorise quadratics.

Example

To multiply out $(2y - 3)(y - 5)$ you should think like this:

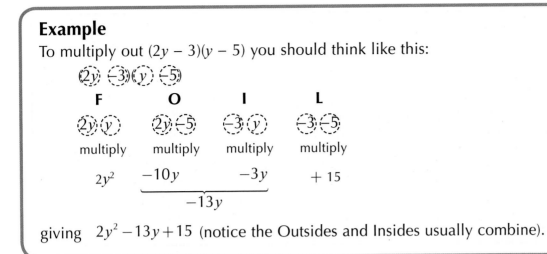

F	O	I	L
multiply	multiply	multiply	multiply
$2y^2$	$-10y$	$-3y$	$+15$

$$-13y$$

giving $2y^2 - 13y + 15$ (notice the Outsides and Insides usually combine).

Removing squared brackets

TOP TIP

$(2x - 3)^2 \; (5x - 4)^2$
Squaring out terms like
these: the constant term
is always POSITIVE.

Remember that m^2 means $m \times m$.

So, for example:

$(3y + 2)^2$ means $(3y + 2)(3y + 2)$

$(b - a)^2$ means $(b - a)(b - a)$

$(5 - x)^2$ means $(5 - x)(5 - x)$

and you then use your FOIL rule to get rid of the two pairs of brackets:

$9y^2 + 6y + 6y + 4 = 9y^2 + 12y + 4$

$b^2 - ab - ab + a^2 = b^2 - 2ab + a^2$

$25 - 5x - 5x + x^2 = 25 - 10x + x^2$

Example

Multiply out the brackets and then simplify: $(3x - 2)^2 - (3x + 2)^2$

Solution

$(3x - 2)(3x - 2) - (3x + 2)(3x + 2)$

$= 9x^2 - 6x - 6x + 4 - (9x^2 + 6x + 6x + 4)^*$

$= 9x^2 - 12x + 4 - (9x^2 + 12x + 4)$

$= 9x^2 - 12x + 4 - 9x^2 - 12x - 4$

$= -24x$

*Note: Brackets are needed because the term $(3x + 2)(3x + 2)$ is being subtracted.

Removing brackets containing more than two terms

For expressions such as:

$$(a + b)(c + d + e)$$

FOIL does not work.

Here is the pattern to use:

$(a + b)(c + d + e)$ giving $ac + ad + ae$

then $(a + b)(c + d + e)$ giving $bc + bd + be$.

There are a total of six multiplications:

$$(a + b)(c + d + e) = ac + ad + ae + bc + bd + be$$

Example

Expand $(5m - 2)(m^2 - 2m + 3)$

Solution

First multiply by $5m$

$(5m - 2)(m^2 - 2m + 3)$ giving $5m^3 - 10m^2 + 15m$

Now multiply by -2

$(5m - 2)(m^2 - 2m + 3)$ giving $-2m^2 + 4m - 6$

So $(5m - 2)(m^2 - 2m + 3)$

$$= 5m^3 - 10m^2 + 15m \quad \text{(from } 5m\text{)}$$

$$-2m^2 + 4m - 6 \quad \text{(from } -2\text{)}$$

$$= 5m^3 - 12m^2 + 19m - 6 \quad \text{(combining 'like' terms)}$$

Quick Test 5

Multiply out and simplify if possible.

1. $7(x - 4)$
2. $-3(2 - 3x)$
3. $-(2x^2 + x - 1)$
4. $(n - 3)(n + 2)$
5. $(3m - 5)(2m + 3)$
6. $(5y - 2)^2$
7. $(4 - 5a)^2 - a(2a - 3)$

Factorising algebraic expressions

What is a factor?

number	factorisations using two factors		factors
30	1×30	2×15	1 2 3 5
	3×10	5×6	6 10 15 30

When 30 is written as a product of two positive integers then each of these integers is a **factor** of 30.

algebraic term	factorisations using two factors		factors
$6n$	$1 \times 6n$	$2 \times 3n$	1 2 3 6
	$3 \times 2n$	$6 \times n$	n $2n$ $3n$ $6n$

When $6n$ is written as a product of two terms then each of these terms is a **factor** of $6n$.

> ## TOP TIP
>
> $8 = 3 + 5$ but 3 and 5 *are not* factors of 8. } Factors come from products
> $8 = 2 \times 4$ and so 2 and 4 } not from sums.
> *are* factors of 8.

Common factors

Here is an example:

$$20ab \quad + \quad 8a \quad = \quad 4a(5b + 2)$$

$4a \times 5b$ $4a \times 2$

The factor $4a$ is common to both terms $20ab$ and $8a$. It is a **common factor**.

The factor $4a$ now appears as one factor in a two factor product $4a \times (5b + 2)$. The other factor is $5b + 2$.

The expression $20ab + 8a$ (a sum of two terms) has been written in factorised form: $4m(5n + 2)$ (a product of two factors).

Examples

Factorise:

1. $9k - 3$ 2. $8mn^2 + 12m^2n$

Solutions

1. $9k - 3 = 3(3k - 1)$
    $(3 \times 3k)$ (3×1)

2. $8mn^2 - 12m^2n = 4mn(2n - 3m)$
    $(4mn \times 2n)$ $(4mn \times 3m)$

> ## TOP TIP
>
> Always take the <u>highest</u> common factor outside the brackets.

What does 'fully factorised' mean?

Here is a factorisation: $6x + 12 = 3(2x + 4)$.

However the factor $2x + 4$ can still be factorised as $2(x + 2)$.

Here is the full factorisation: $6x + 12 = 6(x + 2)$.

This fully factorised form has no further common factor (other than 1!).

Factorising quadratic expressions

You know that $(x-4)(x+3) = x^2 - x - 12$ You expand the pair of brackets using FOIL

$$x^2 \; + \; 3x \; - \; 4x \; - \; 12$$
$$\text{F} \qquad \text{O} \qquad \text{I} \qquad \text{L}$$

Reversing this gives $x^2 - x - 12 = (x-4)(x+3)$ You factorise the quadratic expression by reversing the FOIL expansion but how is this done?

$$(? ? ?)(? ? ?)$$

Vital observation

The **middle term** $-x$ is a combination of the Outsides $(+3x)$ and Insides $(-4x)$.

Example Let's try to factorise $x^2 + 5x - 6$

Step 1 **List all the possibilities for the Firsts and Lasts:**

$(x \quad 2)(x \quad 3)$ The Firsts multiply to give x^2 so $x \times x$.

$(x \quad 1)(x \quad 6)$ The Lasts multiply to give 6 so 2×3 or 1×6.
You ignore $+$ and $-$ signs at this stage.

Step 2 **Use FOIL to find the Outsides and Insides in each case:**

$(x \quad 2)(x \quad 3)$ $3x$ and $2x$ Outsides: $x \times 3$ Insides: $2 \times x$

$(x \quad 1)(x \quad 6)$ $6x$ and x Outsides: $x \times 6$ Insides: $1 \times x$

Step 3 **Attempt to get the middle term $(+5x)$ from the Outsides and Insides:**

$(x \quad 2)(x \quad 3)$ $3x$ and $2x$ $+3x + 2x$ Two $+$ signs

$(x \quad 1)(x \quad 6)$ $6x$ and x $+6x - x$ One $+$ and one $-$ sign

Step 4 **Fill in the $+\!/\!-$ signs and check by expanding out using FOIL:**

$(x+2)(x+3) = x^2 + 3x + 2x + 6$ not correct as -6 is required

$(x-1)(x+6) = x^2 + 6x - x - 6$ correct as this gives $x^2 + 5x - 6$

So $x^2 + 5x - 6 = (x-1)(x+6)$

Example   Factorise $a^2 - 10a + 16$

Working:

Possible Firsts and Lasts	Outsides	Insides	Combine to give $-10a$?
$(a \quad 4)(a \quad 4)$	$4a$	$4a$	not possible
$(a \quad 2)(a \quad 8)$	$8a$	$2a$	$-8a - 2a$ gives $-10a$
$(a \quad 1)(a \quad 16)$	$16a$	a	not possible

Check $(a-2)(a-8) = a^2 - 8a - 2a + 16 = a^2 - 10a + 16$ ✓

More factorising quadratic expressions

Example

To factorise $5x^2 - 29x - 6$ you must be careful to list **all** the possibilities. Since $5x$ and x are not identical Firsts then 'swapping' the Lasts creates different possibilities.

Possible Firsts and Lasts	Outsides	Insides	Combine to give $-29x$?
$(5x \quad 2)(x \quad 3)$	$15x$	$2x$	not possible
$(5x \quad 3)(x \quad 2)$	$10x$	$3x$	not possible
$(5x \quad 1)(x \quad 6)$	$30x$	x	$-30x + x = -29x$
$(5x \quad 6)(x \quad 1)$	$5x$	$6x$	not possible

Check:　　$(5x + 1)(x - 6) = 5x^2 - 30x + x - 6 = 5x^2 - 29x - 6$ ✓

So　　　　$5x^2 - 29x - 6 = (5x + 1)(x - 6)$

TOP TIP

Always take out any common factor first, e.g.
$2a^2 - 2b^2 = 2(a^2 - b^2)$
$= 2(a - b)(a + b)$

Difference of two squares

Choose a pair of square numbers, subtract the smaller from the larger then factorise the result:

$$
\begin{array}{ccc}
49 & - \quad 25 & = \quad 24 \\
(7^2) & (5^2) & (2 \times 12) \\
81 & - \quad 16 & = \quad 65 \\
(9^2) & (4^2) & (5 \times 13)
\end{array}
$$

A pattern can be spotted:

$a^2 - b^2$
$= (a - b)(a + b)$

Square numbers

1 (1^2)	16 (4^2)	100 (10^2)
4 (2^2)	36 (6^2)	64 (8^2)
9 (3^2)	25 (5^2)	
49 (7^2)		81 (9^2)

Check this pattern for more differences of two of the squares above.

Example
Factorise:
$16x^2 - 1$

Solution
$16x^2 - 1 = (4x - 1)(4x + 1)$ (This is $(4x)^2 - 1^2$)

Quick Test 6

1. Factorise:　　a) $6a^2 - 8a$　　　b) $14mn + 7n$
2. Factorise:　　a) $x^2 - 5x - 14$　　b) $3k^2 + 7k - 6$
3. Factorise:　　a) $121 - m^2$　　　b) $18g^2 - 98f^2$

The area of squares

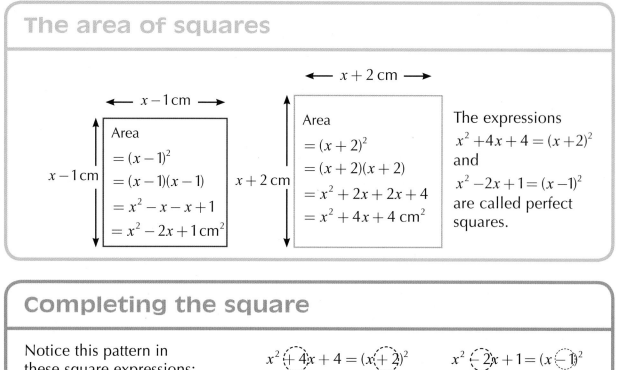

The expressions
$x^2 + 4x + 4 = (x + 2)^2$
and
$x^2 - 2x + 1 = (x - 1)^2$
are called perfect squares.

Completing the square

Notice this pattern in these square expressions:

$x^2 + 4x + 4 = (x + 2)^2$ $x^2 - 2x + 1 = (x - 1)^2$

halve   halve

You can use this pattern to complete expressions to make them squares:

$x^2 - 6x$ becomes $(x - 3)^2 = (x - 3)(x - 3) = x^2 - 6x + 9$

$x^2 + 10x$ becomes $(x + 5)^2 = (x + 5)(x + 5) = x^2 + 10x + 25$

but notice that 9 and 25 have been added to the original expressions.
By now subtracting 9 and 25 the original expressions will have been unaltered:

$x^2 - 6x = (x - 3)^2 - 9$ and $x^2 + 10x = (x + 5)^2 - 25$

Writing a quadratic expression in the form $(x + a)^2 + b$ is called 'completing the square'.

Example
Express in the form $(x + a)^2 + b$
$x^2 + 4x + 1$

Solution
$x^2 + 4x + 1 = (x + 2)(x + 2) - 4 + 1 = (x + 2)^2 - 3$

$x^2 + 4x + 4$ to remove the unwanted +4

TOP TIP

You will use 'completing the square' to help you graph the values of quadratic expressions in Unit 2.

Quick Test 7

Express in the form $(x + a)^2 + b$ where a and b are constants:

1. $x^2 - 12x$ 2. $x^2 + 2x - 3$ 3. $x^2 - 3x + \dfrac{1}{4}$

Simplifying algebraic fractions

Cancelling factors

Numerical

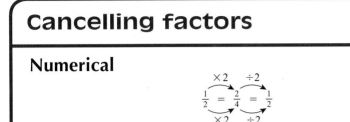

$$\overset{\times 2}{\underset{\times 2}{\frac{1}{2}}} = \overset{}{\underset{}{\frac{2}{4}}} \overset{\div 2}{\underset{\div 2}{= \frac{1}{2}}}$$

Algebraic

$$\overset{\times c}{\underset{\times c}{\frac{a}{b}}} = \overset{}{\underset{}{\frac{ac}{bc}}} \overset{\div c}{\underset{\div c}{= \frac{a}{b}}}$$

> **TOP TIP**
>
> When asked to simplify a fraction you need to cancel any common factor from the top and the bottom.

You can multiply or divide the top (numerator) and bottom (denominator) of a fraction by the same number (or letter) and get an equivalent fraction.

By dividing the top and bottom of this fraction by c: $\frac{ac}{bc} = \frac{a \times c}{b \times c}$ the fraction simplifies to: $\frac{a}{b}$.

The factor c has been cancelled. Usually you score out the factor that you are cancelling: $\frac{a \times \cancel{c}}{b \times \cancel{c}} = \frac{a}{b}$.

Examples

1. $\frac{2a}{4b} = \frac{\cancel{2} \times a}{\cancel{2} \times 2b} = \frac{a}{2b}$ (divide top and bottom by 2)

2. $\frac{3m}{m} = \frac{3 \times \cancel{m}}{1 \times \cancel{m}} = \frac{3}{1} = 3$ (divide top and bottom by m)

3. $\frac{k^2}{5k} = \frac{k \times \cancel{k}}{5 \times \cancel{k}} = \frac{k}{5}$ (divide top and bottom by k)

4. $\frac{(3x-1)^2}{5(3x-1)} = \frac{(3x-1) \times \cancel{(3x-1)}}{5 \times \cancel{(3x-1)}} = \frac{3x-1}{5}$ (divide top and bottom by $3x-1$)

When is cancelling allowed?

Only common **factors** can be cancelled in a fraction.

$\frac{a \times \cancel{c}}{b \times \cancel{c}}$ ←——— These are both multiplications so the common factor c
←——— can be cancelled.

$\frac{a + c}{b + c}$ ←——— These are additions so c is not a factor.
←——— No cancelling is allowed.

If you are not sure whether you are allowed to cancel, try replacing the letters by numbers to see if it works!

Can you cancel here? $\frac{a+b}{a}$ Try $\frac{2+1}{2} \neq 1$. No cancelling allowed.

Can you cancel here? $\frac{a \times a}{a+b}$ Try $\frac{1 \times 1}{1+2} \neq \frac{1}{2}$. No cancelling allowed.

Remember that the letters just stand for unknown numbers.

Factorising then cancelling

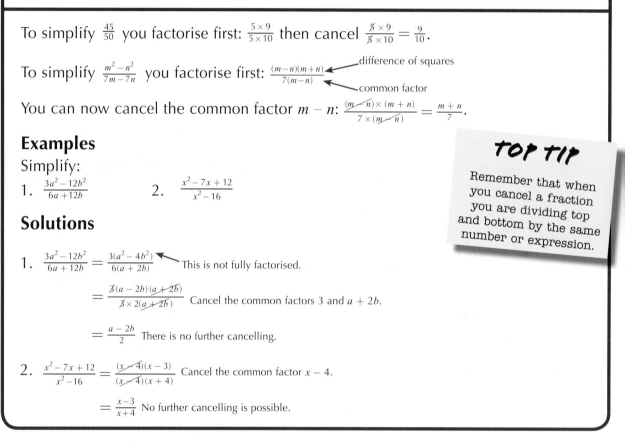

To simplify $\frac{45}{50}$ you factorise first: $\frac{5 \times 9}{5 \times 10}$ then cancel $\frac{\cancel{5} \times 9}{\cancel{5} \times 10} = \frac{9}{10}$.

To simplify $\frac{m^2 - n^2}{7m - 7n}$ you factorise first: $\frac{(m-n)(m+n)}{7(m-n)}$ — difference of squares / common factor

You can now cancel the common factor $m - n$: $\frac{(m - n) \times (m + n)}{7 \times (m - n)} = \frac{m + n}{7}$.

Examples

Simplify:

1. $\frac{3a^2 - 12b^2}{6a + 12b}$ 2. $\frac{x^2 - 7x + 12}{x^2 - 16}$

TOP TIP

Remember that when you cancel a fraction you are dividing top and bottom by the same number or expression.

Solutions

1. $\frac{3a^2 - 12b^2}{6a + 12b} = \frac{3(a^2 - 4b^2)}{6(a + 2b)}$ This is not fully factorised.

   $= \frac{\cancel{3}(a - 2b)(a + 2b)}{\cancel{3} \times 2(a + 2b)}$ Cancel the common factors 3 and $a + 2b$.

   $= \frac{a - 2b}{2}$ There is no further cancelling.

2. $\frac{x^2 - 7x + 12}{x^2 - 16} = \frac{(x - 4)(x - 3)}{(x - 4)(x + 4)}$ Cancel the common factor $x - 4$.

   $= \frac{x - 3}{x + 4}$ No further cancelling is possible.

Quick Test 8

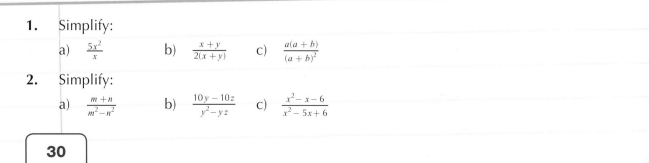

1. Simplify:

   a) $\frac{5x^2}{x}$ b) $\frac{x + y}{2(x + y)}$ c) $\frac{a(a + b)}{(a + b)^2}$

2. Simplify:

   a) $\frac{m + n}{m^2 - n^2}$ b) $\frac{10y - 10z}{y^2 - yz}$ c) $\frac{x^2 - x - 6}{x^2 - 5x + 6}$

Multiplying and dividing algebraic fractions

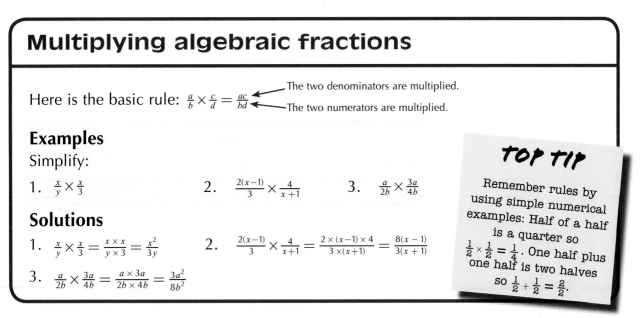

Multiplying algebraic fractions

Here is the basic rule: $\frac{a}{b} \times \frac{c}{d} = \frac{ac}{bd}$ ← The two denominators are multiplied.
← The two numerators are multiplied.

Examples
Simplify:

1. $\frac{x}{y} \times \frac{x}{3}$ 2. $\frac{2(x-1)}{3} \times \frac{4}{x+1}$ 3. $\frac{a}{2b} \times \frac{3a}{4b}$

Solutions

1. $\frac{x}{y} \times \frac{x}{3} = \frac{x \times x}{y \times 3} = \frac{x^2}{3y}$ 2. $\frac{2(x-1)}{3} \times \frac{4}{x+1} = \frac{2 \times (x-1) \times 4}{3 \times (x+1)} = \frac{8(x-1)}{3(x+1)}$

3. $\frac{a}{2b} \times \frac{3a}{4b} = \frac{a \times 3a}{2b \times 4b} = \frac{3a^2}{8b^2}$

TOP TIP

Remember rules by using simple numerical examples: Half of a half is a quarter so $\frac{1}{2} \times \frac{1}{2} = \frac{1}{4}$. One half plus one half is two halves so $\frac{1}{2} + \frac{1}{2} = \frac{2}{2}$.

Special cases

$a \times \frac{b}{c} = \frac{a}{1} \times \frac{b}{c} = \frac{a \times b}{1 \times c} = \frac{ab}{c}$ The factor a appears on the top of the fraction $\frac{b}{c}$.

$\frac{1}{2}m = \frac{1}{2} \times m = \frac{1}{2} \times \frac{m}{1} = \frac{1 \times m}{2 \times 1} = \frac{m}{2}$ This makes sense since finding half of m is the same as dividing m by 2.

Examples $\frac{1}{3}x = \frac{x}{3}$ $\frac{2}{3}y = \frac{2y}{3}$ $\frac{1}{4}x^2 = \frac{x^2}{4}$ $\frac{3}{4}x^4 = \frac{3x^4}{4}$

Multiplying and cancelling

Note: $\frac{a}{c} \times \frac{c}{b} = \frac{a \times \cancel{c}}{\cancel{c} \times b} = \frac{a}{b}$. Dividing top and bottom by c allows you to cancel the common factor.

Alternatively: $\frac{a}{\cancel{c}} \times \frac{\cancel{c}}{b} = \frac{a}{b}$. The cancelling can be done before multiplying.

Examples
Simplify:

1. $\frac{3m}{4} \times \frac{8n}{m}$ 2. $\frac{2}{3(x-1)^2} \times \frac{x-1}{4}$

Solutions

1. $\frac{3m}{4} \times \frac{8n}{m} = \frac{3\not{m}}{\not{4}} \times \frac{\not{4} \times 2n}{\not{m}} = 6n$ The cancelled factors are 4 and m.

2. $\frac{2}{3(x-1)^2} \times \frac{x-1}{4} = \frac{1 \times \not{2}}{3(x-1)(\not{x-1})} \times \frac{(\not{x-1})}{\not{2} \times 2} = \frac{1}{6(x-1)}$ The cancelled factors are 2 and $x-1$.

Dividing algebraic fractions

An efficient and understandable method is the following:

Step 1 Rewrite the division as a 'double-decker' fraction $\frac{a}{b} \div \frac{c}{d}$ becomes $\frac{a/b}{c/d}$.

Step 2 Multiply top and bottom by the same factor to remove the 'double-decker' nature of the fraction $\frac{a/b \times \not{b} \times d}{c/d \times b \times \not{d}} = \frac{ad}{cb}$.

Example

Simplify

1. $\frac{2m}{n} \div \frac{m^2}{n}$

2. $\frac{3k}{7(k+1)} \div \frac{6}{(k+1)^2}$

TOP TIP

Many errors are made by not fully understanding the 'trick' for dividing. 'Turn the second fraction upside down and multiply'. Be very careful that you understand.

Solutions

1. $\frac{2m/n \times \not{n}}{m^2/n \times \not{n}} = \frac{2m}{m^2} = \frac{2\not{m}}{m \times \not{m}} = \frac{2}{m}$

2. $\frac{\frac{3k}{7(k+1)} \times (k+1) \times (k+1) \times 7}{\frac{6}{(k+1)(k+1)} \times (k+1) \times (k+1) \times 7} = \frac{\not{3}k(k+1)}{2 \times \not{3} \times 7} = \frac{k(k+1)}{14}$

Special cases

This method easily copes with examples like these:

$\frac{3x}{4} \div 2$ changes to $\frac{3x/4}{2}$ now multiply top and bottom by 4: $\frac{3x/\not{4} \times \not{4}}{2 \times 4} = \frac{3x}{8}$

$8 \div \frac{a^2}{6}$ changes to $\frac{8}{a^2/6}$ now multiply top and bottom by 6: $\frac{8 \times 6}{a^2/\not{6} \times \not{6}} = \frac{48}{a^2}$

Quick Test 9

1. Simplify

   a) $\frac{2}{n^2} \times \frac{3n^3}{4}$ b) $y \times \frac{2(y-1)}{3y}$ c) $\frac{m}{6} \times \frac{3(m+1)}{m^2}$

2. Express each of these as a single fraction in its simplest form:

   a) $2x \div \frac{2}{3x}$ b) $\frac{1}{a} \div \frac{1}{a^2}$ c) $\frac{2b}{a^2} \div 4b$

GOT IT? ☐ ☐ ☐

Adding and subtracting algebraic fractions

Adding algebraic fractions

Your aim is to get the denominators the same, i.e. to make a **common denominator**.

Numerical

$$\frac{2}{5}+\frac{1}{3}=\frac{2\times3}{5\times3}+\frac{1\times5}{3\times5}=\frac{6}{15}+\frac{5}{15}$$

Fifths and thirds   now both fifteenths

different denominators   common denominators

Algebraic

$$\frac{a}{b}+\frac{c}{d}=\frac{a\times d}{b\times d}+\frac{b\times c}{b\times d}=\frac{ad}{bd}+\frac{bc}{bd}$$

different denominators   common denominators

$$\frac{6}{15}+\frac{5}{15}=\frac{6+5}{15}=\frac{11}{15}$$

you can now add the numerators.

$$\frac{ad}{bd}+\frac{bc}{bd}=\frac{ad+bc}{bd}$$

you can now add the numerators.

Examples

Express these as a single fraction in its simplest form: 1. $\frac{2}{m}+\frac{1}{n}$ 2. $\frac{5}{x}+\frac{2}{x-1}$

Solutions

1. $\frac{2}{m}+\frac{1}{n}=\frac{2\times n}{m\times n}+\frac{m\times1}{m\times n}=\frac{2n}{mn}+\frac{m}{mn}=\frac{2n+m}{mn}$

2. $\frac{5}{x}+\frac{2}{x-1}=\frac{5(x-1)}{x(x-1)}+\frac{2x}{x(x-1)}=\frac{5(x-1)+2x}{x(x-1)}=\frac{5x-5+2x}{x(x-1)}=\frac{7x-5}{x(x-1)}$

Finding the lowest common denominator

When making the denominators the same look closely at the factors of each denominator.

$\frac{2}{a}+\frac{1}{3a^2}$ Here the common denominator will be $3\times a\times a$ so term a should be multiplied by $3a$

$$=\frac{2\times3a}{a\times3a}+\frac{1}{3a^2}=\frac{6a}{3a^2}+\frac{1}{3a^2}=\frac{6a+1}{3a^2}$$

$\frac{3}{2ab}+\frac{2}{3a^2b}$ Here the common denominator will be $2\times3\times a\times a\times b=6a^2b$. $2ab$ is multiplied by $3a$ and $3a^2b$ is multiplied by 2.

$$=\frac{3\times3a}{2ab\times3a}+\frac{2\times2}{3a^2b\times2}=\frac{9a}{6a^2b}+\frac{4}{6a^2b}=\frac{9a+4}{6a^2b}$$

TOP TIP

Always check to see if your final answer can be cancelled down.

Examples

Express these as a single fraction:

1. $\frac{2}{3y^2}+\frac{1}{2y}$ 2. $\frac{5}{m^2n}+\frac{2}{mn^2}$

Solutions

1. $\frac{2}{3y^2}+\frac{1}{2y}=\frac{2\times2}{3y^2\times2}+\frac{1\times3y}{2y\times3y}=\frac{4}{6y^2}+\frac{3y}{6y^2}=\frac{4+3y}{6y^2}$

2. $\frac{5}{m^2n}+\frac{2}{mn^2}=\frac{5\times n}{m^2n\times n}+\frac{2\times m}{mn^2\times m}=\frac{5n}{m^2n^2}+\frac{2m}{m^2n^2}=\frac{5n+2m}{m^2n^2}$

Subtracting algebraic fractions

TOP TIP

Any simple term like a, b, $2x$, y^2 can be written as a fraction: $\frac{a}{1}$, $\frac{b}{1}$, $\frac{2x}{1}$, $\frac{y^2}{1}$.

When you have a common denominator subtract the numerators:

$$\frac{a}{b} - \frac{c}{d} = \frac{a \times d}{b \times d} - \frac{b \times c}{b \times d} = \frac{ad}{bd} - \frac{bc}{bd} = \frac{ad - bc}{bd}$$

Example

Express $\frac{x}{x-1} - \frac{x}{x+1}$ as a single fraction in its simplest form.

Solution

$$\frac{x}{x-1} - \frac{x}{x+1} = \frac{x(x+1)}{(x-1)(x+1)} - \frac{x(x-1)}{(x-1)(x+1)} = \frac{x(x+1) - x(x-1)}{(x-1)(x+1)} = \frac{x^2 + x - x^2 + x}{(x-1)(x+1)} = \frac{2x}{x^2 - 1}$$

Some special cases

To simplify $1 - \frac{1}{a}$ rewrite 1 as $\frac{a}{a}$: $\frac{a}{a} - \frac{1}{a} = \frac{a-1}{a}$

$\left(\text{suppose } a = 5 \text{ then } 1 - \frac{1}{a} = 1 - \frac{1}{5} = \frac{4}{5} \text{ and } \frac{a-1}{a} = \frac{5-1}{5} = \frac{4}{5}\right)$.

To simplify $a + \frac{b}{c}$ remember $a = \frac{a}{1}$ so you get $\frac{a \times c}{1 \times c} + \frac{b}{c} = \frac{ac + b}{c}$.

A further example: $a + 1 + \frac{1}{a} = \frac{a^2}{a} + \frac{a}{a} + \frac{1}{a} = \frac{a^2 + a + 1}{a}$

An application

The resistance (R) of two resistors wired in parallel is given by: $\frac{1}{R} = \frac{1}{R_1} + \frac{1}{R_2}$. Let's add

the fractions: $\frac{1}{R_1} + \frac{1}{R_2} = \frac{1 \times R_2}{R_1 \times R_2} + \frac{R_1 \times 1}{R_1 \times R_2} = \frac{R_2}{R_1 R_2} + \frac{R_1}{R_1 R_2} = \frac{R_1 + R_2}{R_1 R_2}$. So $\frac{1}{R} = \frac{R_1 + R_2}{R_1 R_2}$.

Now 'invert' both fractions: $\frac{R}{1} = \frac{R_1 R_2}{R_1 + R_2}$ or $R = \frac{R_1 R_2}{R_1 + R_2}$, a more useful form for the formula.

Quick Test 10

1. Express these as single fractions:

   a) $5 - \frac{2}{m}$ b) $\frac{1}{x-1} - 1$ c) $\frac{1}{a} + \frac{1}{a^2}$

2. Express as a single fraction in its simplest form:

   a) $\frac{7}{m} + \frac{3}{m+1}$ b) $\frac{1}{2(k-2)} - \frac{1}{k(k-2)}$

The gradient formulae

What is gradient?

Gradient is a number that measures the slope or steepness of a line. Divide the vertical distance by the horizontal distance.

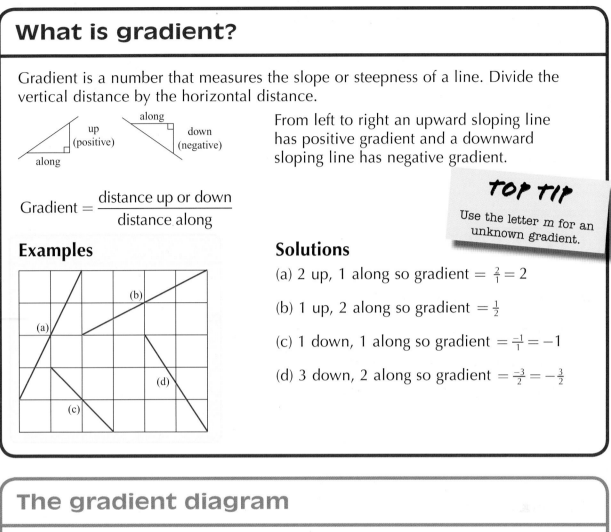

From left to right an upward sloping line has positive gradient and a downward sloping line has negative gradient.

$$\text{Gradient} = \frac{\text{distance up or down}}{\text{distance along}}$$

TOP TIP

Use the letter m for an unknown gradient.

Examples

Solutions

(a) 2 up, 1 along so gradient $= \frac{2}{1} = 2$

(b) 1 up, 2 along so gradient $= \frac{1}{2}$

(c) 1 down, 1 along so gradient $= \frac{-1}{1} = -1$

(d) 3 down, 2 along so gradient $= \frac{-3}{2} = -\frac{3}{2}$

The gradient diagram

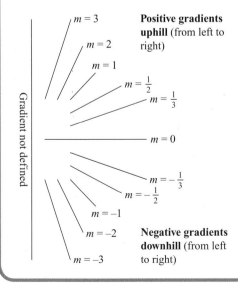

Positive gradients **uphill** (from left to right)

Negative gradients **downhill** (from left to right)

Gradient not defined

Notice a 'horizontal' line has gradient zero whereas a 'vertical' line has no gradient.

You should learn to quickly estimate the gradient of a line just by looking at it.

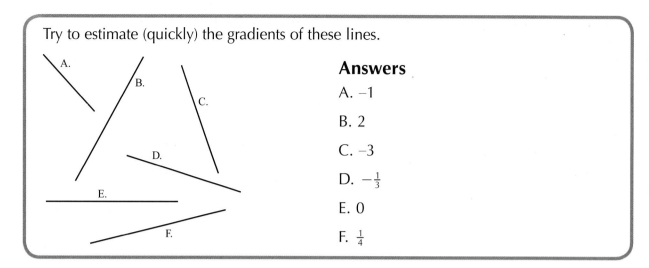

Try to estimate (quickly) the gradients of these lines.

A.
B.
C.
D.
E.
F.

Answers

A. -1

B. 2

C. -3

D. $-\frac{1}{3}$

E. 0

F. $\frac{1}{4}$

The gradient formulae

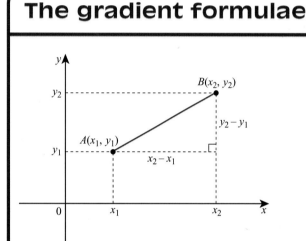

The gradient of line AB is given by:

$$m_{AB} = \frac{y_2 - y_1}{x_2 - x_1}$$ ← The y-coordinate difference.
← The x-coordinate difference.

Note:

1. $\frac{y_1 - y_2}{x_1 - x_2}$ gives the same value (swapping *both* top and bottom subtractions).

2. $x_1 \neq x_2$. If $x_1 = x_2$ then $x_1 - x_2 = 0$ and division by zero is not allowed. In this case AB is 'vertical' and the line has no gradient.

Examples

1. In each case calculate the gradient of the line joining the pair of points
   (a) $A(2, 5)$, $B(6, 13)$ (b) $C(-1, 3)$, $D(-3, 5)$.

2. Show that $m_{PQ} = a + b$ where m_{PQ} is the gradient of the line joining the points $P(a, a^2)$ and $Q(b, b^2)$.

TOP TIP

Start each subtraction using coordinates from the same point.

Solutions

1. (a) $m_{AB} = \frac{13-5}{6-2} = \frac{8}{4} = 2$ (b) $m_{CD} = \frac{5-3}{-3-(-1)} = \frac{2}{-3+1} = \frac{2}{-2} = -1$

2. $m_{PQ} = \frac{b^2-a^2}{b-a}$ $=$ $\frac{(b-a)(b+a)}{b-a}$ $=$ $\frac{\cancel{(b-a)}(b+a)}{\cancel{b-a}}$ $=$ $b+a = a+b$

                                  (difference of   (cancel   (required
                                  two squares)   factor $b-a$)   answer)

Parallel lines

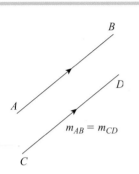

$m_{AB} = m_{CD}$

Parallel lines have the same gradient.

Example
Show that PQ and RS are parallel where
P $(-2, 4)$, Q $(4, 1)$, R $(0, 1)$ and S $(4, -1)$.

Solution
$m_{PQ} = \frac{1-4}{4-(-2)} = \frac{-3}{4+2} = \frac{-3}{6} = -\frac{1}{2}$

$m_{RS} = \frac{-1-1}{4-0} = \frac{-2}{4} = -\frac{1}{2}$, so $m_{PQ} = m_{RS}$

Since the gradients are equal the lines PQ and RS are parallel.

Quick Test 11

1. Calculate the gradient of line AB for:

     a) A $(6, 9)$ and B $(3, 3)$ b) A $(-3, 1)$ and B $(6, -5)$

2. Find the gradient of the line joining P $(m^2, 3m)$ and Q $(n^2, 3n)$ in simplest form.

3. Show that CD and EF are parallel where C $(-1, 1)$, D $(2, 2)$, E $(1, -1)$ and F $(4, 0)$.

Some circle formulae

Circle formulae (a review)

$D = 2r$ Diameter $= 2 \times$ radius

If you take diameter lengths and wrap them around the circumference you find you need just over 3. This number is called Pi and the symbol π is used. $\pi = 3 \cdot 1415926\ldots$ and is an irrational number (the decimal never ends and never repeats).

$C = \pi D$ Circumference $= \pi \times$ Diameter

$A = \pi r^2$ Area $= \pi \times$ radius $\times$ radius

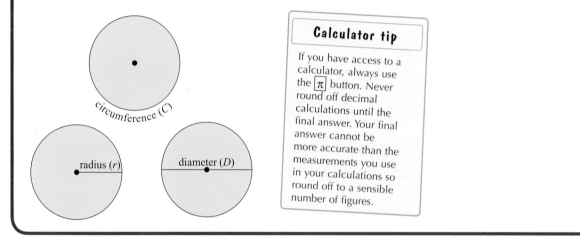

circumference (C)

radius (r)

diameter (D)

> **Calculator tip**
>
> If you have access to a calculator, always use the $\boxed{\pi}$ button. Never round off decimal calculations until the final answer. Your final answer cannot be more accurate than the measurements you use in your calculations so round off to a sensible number of figures.

Arcs and sectors

Arc AB is part of the circumference of the circle.

A

B

O

A circle with centre O:

O

A

B

O

Sector OAB is a 'slice' of the circle.

Finding arcs

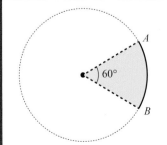

A

$60°$

B

In this case the 'fraction of the circle' is $\frac{60}{360} = \frac{1}{6}$ since a complete turn at the centre is $360°$. So arc $AB = \frac{1}{6} \times \pi D$.

In general if the angle at the centre is $x°$ then the 'circle fraction' will be $\frac{x}{360}$ and arc $AB = \dfrac{x}{360} \times \pi D$.

Remember: $C = \pi D$ or $C = \pi \times 2r = 2\pi r$ (this is the full circumference).

Finding sectors

This time you have to find a fraction of the area of the circle.

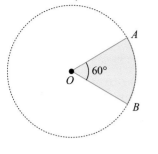

Remember: $A = \pi r^2$

And in this case the 'circle fraction' is $\frac{60}{360} = \frac{1}{6}$ so sector $AOB = \frac{1}{6} \times \pi r^2$.

In general, with $x°$ at the centre, sector $AOB = \frac{x}{360} \times \pi r^2$.

Perimeters of sectors

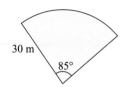

The diagram shows a field in the shape of a sector.

What length of fence is required for this field?

Two straight sections $= 2 \times 30 = 60$ m $\ (2 \times \text{radius})$

Curved section (arc) $= \frac{85}{360} \times \pi \times 60$ (diameter $= 60$ m)

$= 44 \cdot 505 \ldots$ m

Total length $= 44 \cdot 505\ldots + 60 = 104 \cdot 505\ldots \doteqdot 105$ m
(to the nearest m)

Note: perimeter of sector $=$ arc $+ 2 \times$ radius.

TOP TIP

If measurements are to 2 significant figures then your answer cannot be more accurate.

Quick Test 12

1. Find the length of arc PQ:

   a)

   b)

2. Find the area of sector AOB:

   a)

   b)

3. Find the perimeter of these sectors:

   a)

   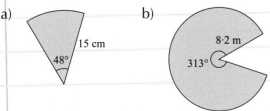

   b)

Some volumes of solids formulae

Areas and volumes (a review)

Area is measured in square units.

Here is a square centimetre:

1 cm | 1 cm² |
1 cm

Volume is measured in cube units.

Here is a centimetre cube:

1 cm
1 cm   1 cm

$1 cm^3$ is also called: 1 cubic centimetre (1 cc)

1 millilitre (1 ml)

$1000 cm^3 = 1$ litre

Basic area formulae:

Rectangle
b | $A = \ell \times b$
ℓ

Triangle
h $A = \dfrac{1}{2}b \times h$
b

Volume of a sphere

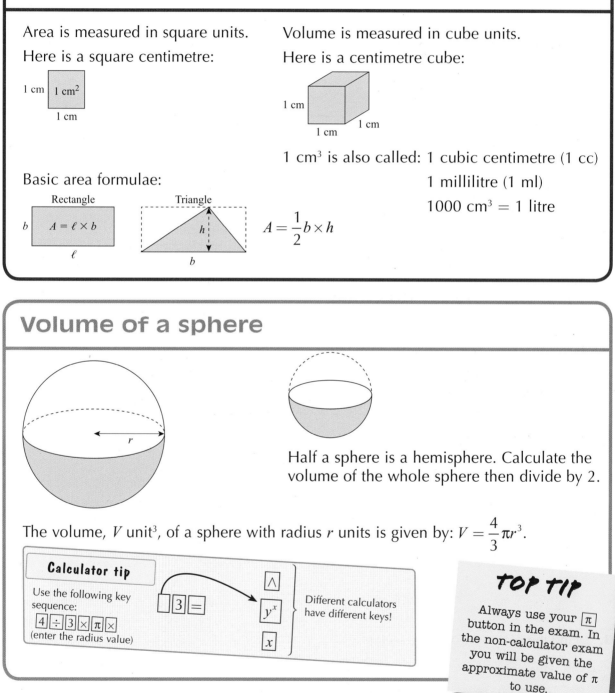

Half a sphere is a hemisphere. Calculate the volume of the whole sphere then divide by 2.

The volume, V unit^3, of a sphere with radius r units is given by: $V = \dfrac{4}{3}\pi r^3$.

Calculator tip

Use the following key sequence:
4 ÷ 3 × π ×
(enter the radius value)

3 =
∧
y^x
x

Different calculators have different keys!

TOP TIP

Always use your π button in the exam. In the non-calculator exam you will be given the approximate value of π to use.

Example

Find the volume of a sphere with diameter 14 cm giving your answer correct to 3 s.f.

Solution

Use $V = \frac{4}{3}\pi r^3$ with $r = 7$ cm (half the diameter)

So $V = \frac{4}{3} \times \pi \times 7^3 = 1436 \cdot 75... \doteq 1440$ cm^3 (to 3 s.f.)

Volume of a cone

The volume, V unit^3, of a cone with perpendicular height h units and radius of base r units is given by: $V = \frac{1}{3}\pi r^2 h$.

Example

An ice cream cone is 12 cm long with diameter 7 cm at the top. How many 1-litre tubs of ice cream are required to fill 150 of these cones?

Solution

Use $V = \frac{1}{3}\pi r^2 h$ with $r = 3 \cdot 5$ cm (half diameter) and $h = 12$ cm

So $V = \frac{1}{3}\pi \times 3 \cdot 5^2 \times 12 = 153 \cdot 938...$ cm^3

150 cones have volume $= 153 \cdot 938... \times 150 = 23\,090 \cdot 7...$ cm^3

now 1000 cm^3 = 1 litre so this is $23 \cdot 09...$ litres. This means 24 tubs are needed.

Notice that 23 tubs would not supply quite enough ice cream!

TOP TIP

If you are asked for litres remember: 1000 cm^3 = 1 litre.

Volume of a pyramid

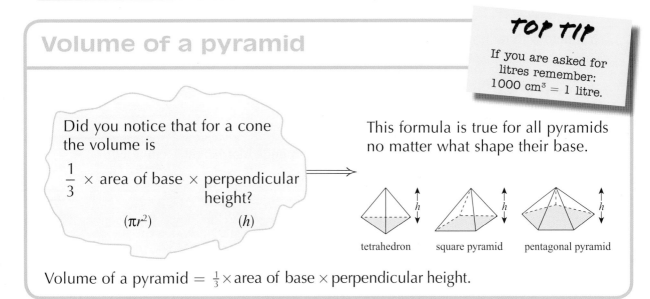

Did you notice that for a cone the volume is

$\frac{1}{3} \times$ area of base $\times$ perpendicular height?

(πr^2) (h)

This formula is true for all pyramids no matter what shape their base.

tetrahedron   square pyramid   pentagonal pyramid

Volume of a pyramid $= \frac{1}{3} \times$ area of base $\times$ perpendicular height.

Unit 1: Expressions and formulae

Volume of a prism

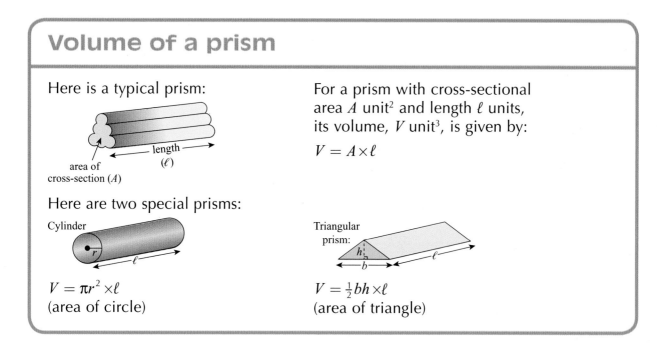

Here is a typical prism:

For a prism with cross-sectional area A unit^2 and length ℓ units, its volume, V unit^3, is given by:

$$V = A \times \ell$$

Here are two special prisms:

Cylinder

$V = \pi r^2 \times \ell$
(area of circle)

Triangular prism:

$V = \frac{1}{2}bh \times \ell$
(area of triangle)

Quick Test 13

1. Find the volume, to 3 s.f., of each of these solids:

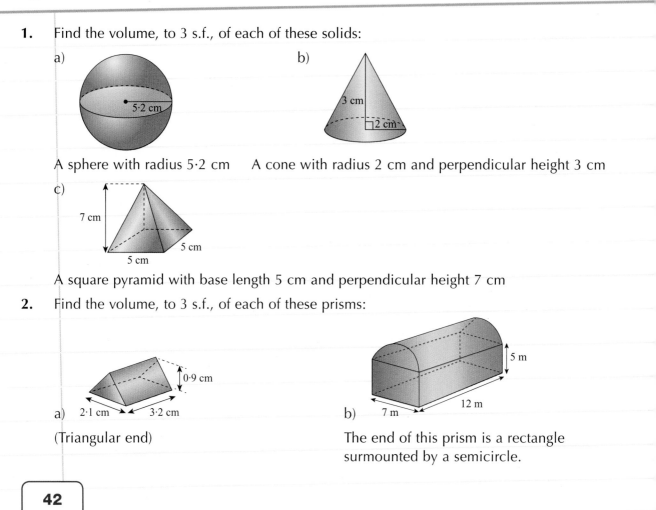

a)

A sphere with radius 5·2 cm

b)

3 cm

2 cm

A cone with radius 2 cm and perpendicular height 3 cm

c)

7 cm

5 cm

5 cm

A square pyramid with base length 5 cm and perpendicular height 7 cm

2. Find the volume, to 3 s.f., of each of these prisms:

a)   2·1 cm   3·2 cm   0·9 cm

(Triangular end)

b)   7 m   12 m   5 m

The end of this prism is a rectangle surmounted by a semicircle.

Sample unit 1 test questions

Surds and indices

1. Simplify $\sqrt{192}$ giving your answer in surd form.

2. Simplify:

   (a) $x^{-2} \times x^{\frac{1}{2}}$

   (b) $\dfrac{5m^8 \times 2m^3}{m^4}$

3. The diameter of a Hydrogen atom is $1 \cdot 06 \times 10^{-10}$ m

   The diameter of its nucleus is 44166 times smaller.

   Calculate the diameter of the nucleus of a Hydrogen atom giving your answer in scientific notation.

Expressions

1. Expand $a(b - 5a)$.

2. Expand and simplify $(x - 2)(x + 7)$.

3. Factorise:

   (a) $6m^2 - m$

   (b) $49 - k^2$

   (c) $p^2 + 9p + 14$

4. Express $x^2 - 2x + 5$ in the form $(x - p)^2 + q$.

Algebraic fractions

1. Write $\dfrac{(a + 2b)^2}{(a + 2b)(a - b)}$ in its simplest form.

2. Express each of these as a single fraction.

   (a) $2 \div \dfrac{3}{x}$

   (b) $\dfrac{1}{m} - \dfrac{2}{n}$

Formulae

1. P is the point $(5, -1)$ and Q is the point $(-1, 4)$. Find the gradient of the line PQ.

2.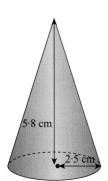

   Calculate the volume of this cone, which has a height of $5 \cdot 8$ cm and a base radius of $2 \cdot 5$ cm.

   Give your answer correct to two significant figures.

3. The diagram shows a discus circle on a sports field. The circle has a radius of $1 \cdot 25$ m. C is the centre, with angle MCN $= 40°$. Minor arc MN is to be marked with black tape.

   (a) Calculate the length of black tape required.

   (b) How many such circles can be marked using a 5 metre roll of tape?

   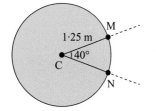

4. A sphere of radius 5 cm is made from plasticine. It is reshaped to make a cylinder with radius 4 cm and height 11 cm. Was more plasticine needed or was there some left over? (Show your reasoning.)

   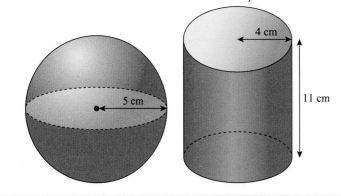

Sample end-of-course exam questions on unit 1 topics

Non-calculator questions

1. Multiply out the brackets and collect like terms: $(3a - 1)(a^2 + 2a - 3)$.

2. Express $\dfrac{5}{x+1} - \dfrac{2}{x}$ as a single fraction.

3. (a) Simplify $\sqrt{125} - \sqrt{45}$.   (b) Evaluate $27^{-\frac{2}{3}}$

   (c) Express $\frac{3}{\sqrt{6}}$ with a rational denominator.

4. (a) Factorise fully $50x^2 - 32$

   (b) Hence simplify $\frac{5x + 4}{50x^2 - 32}$

5. (a) Express $x^{\frac{1}{3}}(x^{-\frac{4}{3}} - x^{-\frac{1}{3}})$ without brackets in its simplest form.

   (b) When $x = 2\frac{1}{2}$ find the exact value of this expression.

6. The sequence 1, 5, 6, 11, 17, ... has the rule that after the first two terms each term is the sum of the previous two terms. The sequence m, n, $m + n$, ... has the same rule.

   (a) Show that t_5, the 5th term, is $3m + 5n$.

   (b) Show that $t_6 = 3t_3 + 2t_2$.

7. $A(-13, -3)$, $B(-1, 5)$ and $C(-5, 2)$ are the three vertices of triangle ABC. Which of the three sides of the triangle has the greatest gradient?

8. An 8 cm length of wire is shaped into a circle.

   (a) Show that the radius of the circle is $\frac{4}{\pi}$ cm.

   (b) Show that the area of the circle is $\frac{16}{\pi}$ cm².

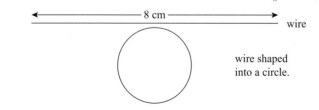

         8 cm   wire

wire shaped into a circle.

Calculator-allowed questions

1. The volume, V, of a sphere is given by $V = \frac{4}{3}\pi r^3$ where r is the radius of the sphere. Taking the radius of the earth to be $6\cdot4\times10^3$ km, calculate its volume in km³ giving your answer in scientific notation correct to 2 significant figures.

2. The distance from Earth to the star Sirius is $8\cdot12\times10^{13}$ km. The speed of light is $3\cdot00\times10^5$ km per second. How long does it take light to travel from Sirius to Earth? Give your answer in years correct to 1 decimal place.

3. The diagram shows a pencil sharpener which takes the form of a hollow cuboid with a cone inside. The circular opening meets three of the edges of the end of the sharpener. The vertex A of the cone lies on the other end of the sharpener and lies at equal distance from three edges of the end, as shown.

   (a) Calculate the volume of the cone.

   (b) Calculate the volume available for pencil shavings.

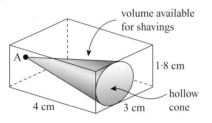

4. The diagram shows a cut-out template to make a paper witch's hat.

   (a) For the brim a 29·4 cm diameter circle is cut out of a 54·6 cm diameter circle. Calculate the area of the brim.

   (b) For the cone a sector of a 36 cm radius circle is used with the angle at the centre being 147°, as shown. Show that the area of the cone sector is equal to the area of the brim.

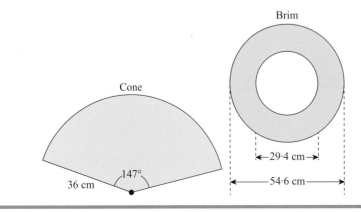

Working with functions

What is a function?

A function describes the relationship between two sets of quantities where one set depends on the other set.

The letters f, g and h are usually used for the names of functions.

Think of a function as:

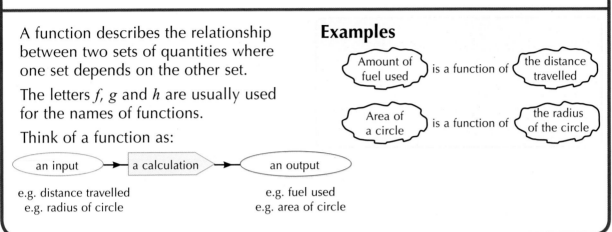

an input → a calculation → an output

e.g. distance travelled
e.g. radius of circle

e.g. fuel used
e.g. area of circle

Examples

Amount of fuel used   is a function of   the distance travelled

Area of a circle   is a function of   the radius of the circle

Function notation

Here is an example of function notation and its meaning:

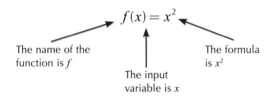

$$f(x) = x^2$$

The name of the function is f

The input variable is x

The formula is x^2

You may evaluate this function for a particular value of x:

$f(5) = 5^2 = 25$.

Every occurrence of x is replaced by 5.

The value of f when $x = 5$ is 25.

An alternative notation is:

$f : x \rightarrow x^2$

$f : 5 \rightarrow 5^2$

The input is 5 and the output is 25.

Example

Given that $g(n) = n^2 - 4n$

evaluate $g(-2)$

Solution

$g(-2) = (-2)^2 - 4 \times (-2) = 4 + 8 = 12$

Graph of a function

Example

Draw the graph of $y = f(x)$ where $f(x) = 2x$.

TOP TIP

To graph a function, e.g. $f(x) = x^2$ replace $f(x)$ by y: $y = x^2$. This gives the equation of the graph.

Solution

Typical points on this graph
are $(-1, -2)$, $(0, 0)$, $(1, 2)$, $(2, 4)$, etc.
where the y-coordinate (output)
is twice the x-coordinate (input).

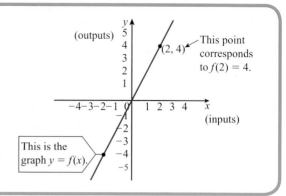

(outputs)

(2, 4) This point corresponds to $f(2) = 4$.

(inputs)

This is the graph $y = f(x)$.

Types of functions

Name:	Formula ($f(x)$)	Typical graphs ($y = f(x)$)
Constant functions	$f(x) = a$ (a is a constant fixed number)	*straight line graphs parallel to the x-axis*
Linear functions	$f(x) = ax + b$ (there is an 'x-term' and possibly a constant term)	*straight line graphs with various non-zero gradients*
Quadratic functions	$f(x) = ax^2 + bx + c$ (there is an 'x²-term', possibly an 'x-term' and possibly a 'constant' term)	*graphs are parabolas*

There are many other types of functions, e.g. trigonometric (trig) functions,
$f(x) = \sin x$ or $\cos x$ or $\tan x$, etc.

Quick Test 14

1. Evaluate $g(-1)$ where $g(x) = 7 - 4x$.

2. Given that $f(n) = n^2 - 5n$ evaluate:

    a) $f(-2)$ b) $f\left(\frac{1}{2}\right)$

3. If $f(x) = x^2$, find two values for x so that $f(x) = 49$.

Working with linear equations and inequations

TOP TIP

It's bad form to write your solution as $7 = x$. Change it to $x = 7$.

Solving linear equations

An equation looks like this: (terms on the left-hand side) = (terms on the right-hand side)

A linear equation has only x-terms and constants. You have solved the equation when you have changed it so that the letter is on its own on one side and there is only a number on the other side, e.g. $x = $ a number or a number $= x$.

You change the equation by doing the same operation to both sides: keep the equation balanced!

Example   Solve $5(x + 5) - 2x = x + 7$

Step 1   Remove any brackets
$5x + 25 - 2x = x + 7$

Step 2   Simplify each side if possible
$3x + 25 = x + 7$

Step 3   Carry out balancing operations

$(-x) \qquad (-x)$ remove x from both sides

$2x + 25 = 7$

$(-25) \quad (-25)$ take away 25 from both sides

$2x = -18$ divide both sides by 2

$(\div 2) \quad (\div 2)$

$x = \dfrac{-18}{2}$

Step 4   State clearly the solution
$x = -9$

Step 5   Check the solution
substituting $x = -9$ in both sides gives $-2 = -2$

Checking your solution

You should always check your solution to an equation.

Kim and Zoe tried to solve
$2(x + 5) = 7x$.

Kim got $x = 3$.

Here is her check:

$2(x + 5) = 2(3 + 5) = 2 \times 8 = 16$
$7x = 7 \times 3 = 21$

Not equal! $x = 3$ is wrong!

Zoe got $x = 2$.

Here is her check:

$2(x + 5) = 2(2 + 5) = 2 \times 7 = 14$
$7x = 7 \times 2 = 14$

Equal! So $x = 2$ is the correct solution.

Solving linear inequations

TOP TIP
Start your solution with the variable, e.g. $x \geq 5$ not $5 \leq x$.

The inequality signs

$<$ is less than $-3 < -2$

$\leq$ is less than or equal to $1 \leq 1$

$>$ is greater than $3 > 2$ $2 > -6$

$\geq$ is greater than or equal to $4 \geq 3$

Solve an inequation as you would an equation.

Apart from: when you multiply or divide both sides by a negative number you switch the inequality sign round, e.g. change $<$ to $>$ or change $>$ to $<$ or change $\leq$ to $\geq$ or change $\geq$ to $\leq$. e.g.

$$-4 < -2 \atop (\div -2)(\div -2)$$

$$\frac{-4}{-2} > \frac{-2}{-2}$$

$$2 > 1$$

$$-3x > -2 \atop (\div -3)(\div -3)$$

$$\frac{-3x}{-3} < \frac{-2}{-3}$$

$$x < \frac{2}{3}$$

- By multiplying both sides by –1:

  $-x < a$ becomes $x > -a$

  $-x > a$ becomes $x < -a$

  $-x < -a$ becomes $x > a$

  $-x > -a$ becomes $x < a$

- $5 > 2$ can be written $2 < 5$

  So $a > x$ can be written $x < a$

  and $a < x$ can be written $x > a$

Example Solve $x + 6 \leq 5x + 10$

Method 1

$$x + 6 \leq 5x + 10 \atop (-x) \qquad (-x)$$

$$6 \leq 4x + 10 \atop (-10) \qquad (-10)$$

$$-4 \leq 4x \atop (\div 4) \quad (\div 4)$$

$$-1 \leq x$$

rewrite as $x \geq -1$

Method 2

$$x + 6 \leq 5x + 10 \atop (-5x) \qquad (-5x)$$

$$-4x + 6 \leq 10 \atop (-6) \qquad (-6)$$

$$-4x \leq 4 \atop (\div -4) \quad (\div -4)$$

switch the sign $x \geq \dfrac{4}{-4}$

$$x \geq -1$$

Note: $x \geq -1$ means all numbers greater than or equal to –1 satisfy the original inequation.

Quick Test 15

1. Solve these equations: a) $3x = 4 - 5x$ b) $2y - 2(3 - 2y) = 3$ c) $-3(x - 3) = 2(5 - x)$

2. Kim and Mia buy fruit at the market. Write down algebraic equations to illustrate:

   a) Kim buys an apple and an orange for 75 pence.

   b) Mia buys two apples and three oranges for £1·95.

3. Solve these inequations: a) $-x > 3$ b) $5 < 1 + x$ c) $2(x - 1) \leq 5(2 + x)$

Graphs from equations

Point plotting

To draw the graph of a line from its equation one method is to plot a few points on the graph and then draw a line through them.

Step 1 Choose at least 3 values for x then use the equation to calculate the corresponding values of y. It is useful to put the values in a table.

Step 2 Each pair of values in your table will be a point on the graph. Draw a coordinate diagram choosing a scale suitable for plotting the points you have in your table.

Step 3 Plot the points on your diagram and draw a line through them. Remember to give your graph a title and to label the line with its equation.

Example Draw the graph $y = 2x + 1$.

Solution

Here is a table of values:

x	−2	−1	0	1	2
y	−3	−1	1	3	5

x-scale: −2 to 2

y-scale: −3 to 5

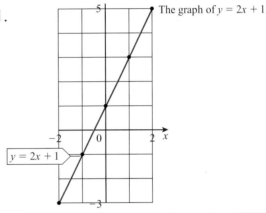

The graph of $y = 2x + 1$

$y = 2x + 1$

Using the gradient and y-intercept

TOP TIP
To find the x-intercept set $y = 0$ in the equation.

When the equation of a straight line is written in the form $y = mx + c$ you know the gradient of the line is m and the point where it crosses the y-axis is $(0, c)$ i.e. the y-intercept is c.

Step 1 If necessary, rearrange the equation into the form:

$y = mx + c$

Step 2 Note the gradient m and the y-intercept c

Step 3 Use this information to sketch the graph. (The gradient diagram on page 35 will help you.)

Example Sketch the graph $y + x = 4$.

Solution Subtracting x from both sides of the equation gives: $y = -x + 4$.

The gradient is −1, the y-intercept is 4.

The graph

$x + y = 4$

Note: Remember that lines with negative gradients slope downwards.

Rearranging equations

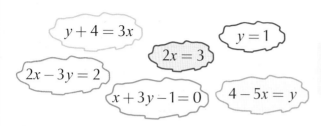

All of these equations have:

- an x-term and/or a y-term
- a constant term
- no other type of term.

They are linear equations with straight line graphs.

All linear equations with a y-term can be rearranged to the form: $y = mx + c$.

Example

Find the gradient of the line $2y + x = 6$.

Solution

$$2y + \underset{(-x)}{x} = \underset{(-x)}{6} \qquad \text{subtract } x \text{ from both sides}$$

$$\underset{(\div 2)}{2y} = \underset{(\div 2)}{-x + 6} \qquad \text{divide both sides (all terms) by 2}$$

$$y = -\tfrac{1}{2}x + 3$$

The gradient is $-\tfrac{1}{2}$.

Special lines

TOP TIP

Lines with equations $y = mx$ all pass through the origin.

1.

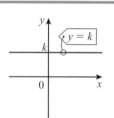

Equations of lines parallel to the x-axis are of the form

$y = $ 'a number'.

2.

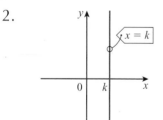

Equations of lines parallel to the y-axis are of the form

$x = $ 'a number'.

3.

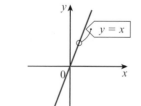

Equation of the x-axis is $y = 0$.
Equation of the y-axis is $x = 0$.

4.

All lines passing through the origin (apart from the y-axis) have equations of the form $y = mx$ where m is the gradient.

Quick Test 16

1. Draw the graphs: a) $y = \tfrac{1}{2}x - 2$ b) $y = -3x + 6$

2. Find the gradient of these lines: a) $2y + 4x = 1$ b) $3y - x - 3 = 0$

3. Sketch these graphs: a) $y - x = 0$ b) $2y - 3x = 0$

Equations from graphs

TOP TIP
Read the scales on both axes carefully.

Working from a diagram

To find the equation of a line from a diagram showing the line graph:

Step 1　Calculate the gradient of the line. To do this you will need to draw a triangle on the grid with the sloping side along the line. Now use:

$$\text{Gradient} = \frac{\text{distance up or down}}{\text{distance along}}$$

Step 2　Find the y-intercept: note the number on the y-axis where the line crosses.

The equation is:

Step 3　　$y = \bigcirc x + \bigcirc$

The number from step 1 (gradient)

The number from step 2 (y-intercept)

Note: You are using '$y = mx + c$' (see page 51).

Examples

Find the equations of these lines:

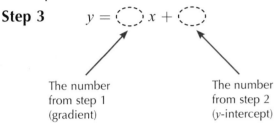

Solutions

(a)　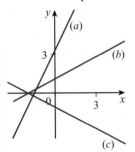 gradient $= \frac{2}{1} = 2$　y-intercept gives 3　equation: $y = 2x + 3$

(b)　gradient $= \frac{1}{2}$　y-intercept gives 1　equation: $y = \frac{1}{2}x + 1$

(c)　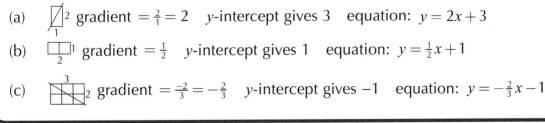 gradient $= \frac{-2}{3} = -\frac{2}{3}$　y-intercept gives -1　equation: $y = -\frac{2}{3}x - 1$

Working from the gradient and y-intercept

If you know the gradient of a line then this is the value of m.

If you know the y-intercept for the line then this is the value of c.

The equation of the line is $y = mx + c$.

Example

Find the equation of the line with gradient 3 which crosses the y-axis at the point $(0, -2)$.

Solution

gradient: $m = 3$

y-intercept: $c = -2$

Equation is $y = 3x - 2$

TOP TIP

Curious fact
Why the letter m for gradient?
There are many theories but nobody really knows!

Points and equations

The coordinates of points on the line satisfy the equation of the line.

The coordinates of points not on the line don't satisfy the equation of the line.

Example

Which of A $(-1, -5)$, B $(-3, 2)$ and C $(-2, 7)$ lies on the line $x + y = 5$?

Solution

For A: $x = -1$, $y = -5$ and $x + y = -1 + (-5) = -6$

For B: $x = -3$, $y = 2$ and $x + y = -3 + 2 = -1$

For C: $x = -2$, $y = 7$ and $x + y = -2 + 7 = 5$

Only C has coordinates that satisfy the equation so it lies on the line.

Working from the gradient and a point

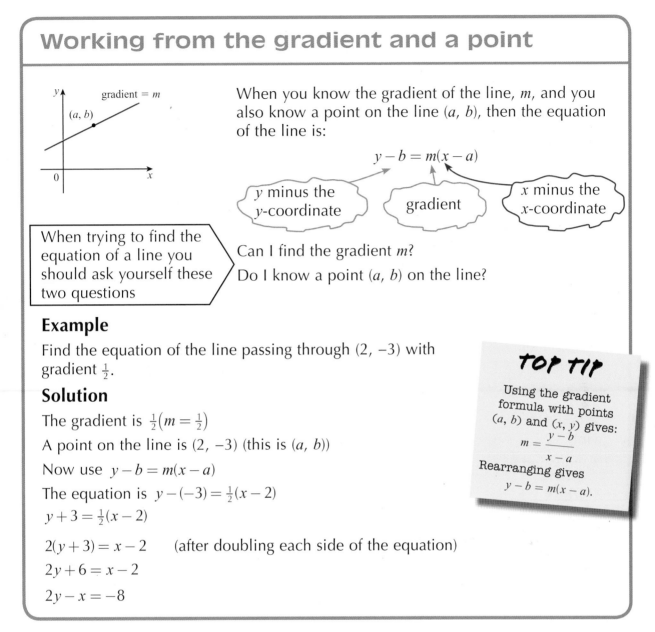

When you know the gradient of the line, m, and you also know a point on the line (a, b), then the equation of the line is:

$$y - b = m(x - a)$$

y minus the *y*-coordinate   gradient   *x* minus the *x*-coordinate

When trying to find the equation of a line you should ask yourself these two questions

Can I find the gradient m?

Do I know a point (a, b) on the line?

Example

Find the equation of the line passing through $(2, -3)$ with gradient $\frac{1}{2}$.

Solution

The gradient is $\frac{1}{2}$ $(m = \frac{1}{2})$

A point on the line is $(2, -3)$ (this is (a, b))

Now use $y - b = m(x - a)$

The equation is $y - (-3) = \frac{1}{2}(x - 2)$

$y + 3 = \frac{1}{2}(x - 2)$

$2(y + 3) = x - 2$ (after doubling each side of the equation)

$2y + 6 = x - 2$

$2y - x = -8$

TOP TIP

Using the gradient formula with points (a, b) and (x, y) gives:

$$m = \frac{y - b}{x - a}$$

Rearranging gives

$$y - b = m(x - a).$$

Quick Test 17

1. Find the equations for these straight line graphs:

a)

b)   A$(-6, 3)$ B$(12, 9)$

2. Line 1: $y = 2x + 3$, Line 2: $2y - x = 3$. For each line determine which of A$(1, 5)$, B$(1, 2)$ and C$(-1, 1)$ lies on the line.

3. Find the equation of the line passing through P$(-2, 5)$ and Q$(-1, 9)$.

Working with linear graphs

Gradient and point (alternative method)

When you know the gradient, m, of a line and a point (a, b) on the line then you have seen that the equation is

$y - b = m(x - a)$.

However it is still possible to use

$y = mx + c$

in this case.

Suppose, for instance, that you know $m = 3$, then using $y = mx + c$ the equation is

$y = 3x + c$.

Since the coordinate values of any point on this line will satisfy its equation (see page 36) then you can use $x = a$ and $y = b$ from (a, b) by substituting the values in the equation.

Suppose you now know $(1, 2)$ lies on the line, then use $x = 1$ and $y = 2$ in $y = 3x + c$:

$\qquad 2 = 3 \times 1 + c$.

giving $\quad 2 = 3 + c \quad$ so $\quad c = -1$.

The equation is $y = 3x - 1$.

Example

Find the equation of the line with gradient -2 which passes through the point $(2, -6)$.

Solution

Using $y = mx + c$ with $m = -2$ the equation is

$y = -2x + c$

$(2, -6)$ lies on the line so substitute $x = 2$ and $y = -6$ in the equation:

$-6 = -2 \times 2 + c$

so $-6 = -4 + c$

so $c = -2$.

The equation is $y = -2x - 2$.

Note: Using $y - b = m(x - a)$

with $m = -2$ and (a, b) being $(2, -6)$

gives $y - (-6) = -2(x - 2)$

so $y + 6 = -2x + 4$

so $y = -2x - 2$ (as above).

TOP TIP

When you have a choice of methods choose the one you find the easiest to use.

Working with different variable names

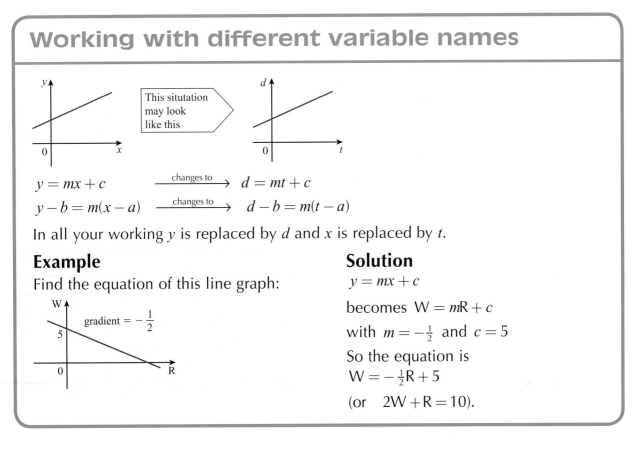

$$y = mx + c \xrightarrow{\text{changes to}} d = mt + c$$
$$y - b = m(x - a) \xrightarrow{\text{changes to}} d - b = m(t - a)$$

In all your working y is replaced by d and x is replaced by t.

Example
Find the equation of this line graph:

Solution
$y = mx + c$

becomes $W = mR + c$

with $m = -\frac{1}{2}$ and $c = 5$

So the equation is
$W = -\frac{1}{2}R + 5$

(or $\quad 2W + R = 10$).

Working with contexts

Example
The graph shows the distance (d km) to Edinburgh against time (t hours) from the start of a train's journey.

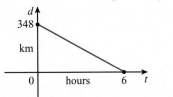

(a) Find the equation of the line in terms of d and t.

(b) Find the distance remaining after 5 hours.

Solution
(a) Points on the graph are (0, 348), (6, 0) so gradient $= \dfrac{348 - 0}{0 - 6} = \dfrac{348}{-6} = -58$

   d-axis intercept is 348 so the equation is $d = -58t + 348$

   (using '$y = mx + c$')

(b) Substitute $t = 5$ in the equation:
   $d = -58 \times 5 + 348 = -290 + 348 = 58$

So after 5 hours there are 58 km to go.

Example

An internet provider charges for the quantity (Q MBs) downloaded.

Here is a graph showing the charges:

If this is a linear graph calculate how many MB can be downloaded for a charge of 80p.

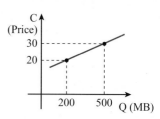

Solution

Points on the graph are (200, 20) and (500, 30).

So gradient $= \dfrac{30-20}{500-200} = \dfrac{10}{300} = \dfrac{1}{30}$.

A point on the graph is (200, 20) so the equation is

$C - 20 = \dfrac{1}{30}(Q - 200)$ (using '$y - b = m(x - a)$')

so $30C - 600 = Q - 200$ (multiplying by 30)

so $30C = Q + 400$ (adding 600 to both sides).

Now let $C = 80$ so $30 \times 80 = Q + 400$,

giving $2400 = Q + 400$ so $Q = 2000$.

2000 MB may be downloaded for 80p.

Quick Test 18

1. Using two different methods, find the equation of the line with gradient $\frac{1}{2}$ that passes through the point (2, −3).

2. Find the equation of each of these line graphs:

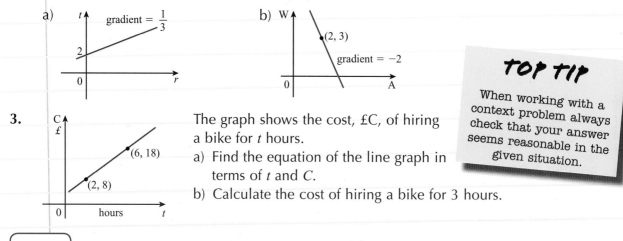

a) gradient $= \frac{1}{3}$

b) (2, 3) gradient $= -2$

3.

The graph shows the cost, £C, of hiring a bike for t hours.

a) Find the equation of the line graph in terms of t and C.

b) Calculate the cost of hiring a bike for 3 hours.

TOP TIP

When working with a context problem always check that your answer seems reasonable in the given situation.

Working with simultaneous equations graphically

What are simultaneous equations?

If $y = 2x$ then there are many pairs of values for x and y that satisfy this equation, e.g. $x = 0$, $y = 0$ and $x = 1$, $y = 2$ etc.

These can be shown on a graph as points, e.g. (0, 0), (1, 2) etc.

Similarly for equation $y = x + 3$, e.g. (0, 3), (1, 4) etc.

When both equations must be satisfied then they are called **simultaneous equations**.

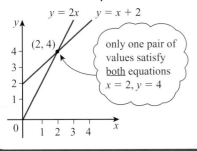

only one pair of values satisfy <u>both</u> equations
$x = 2$, $y = 4$

Setting up a linear equation

Two adults and three children go to the cinema. Their tickets cost £13 in total. If the price of an adult and a child's ticket are not known then letters are used:

Child's ticket: £x Adult's ticket: £y

Total cost is £$(3x + 2y)$

3 lots of £x 2 lots of £y

So $3x + 2y = 13$ (a linear equation)

Example

The total cost of my journey one day was £15. The bus cost 30p per km and the taxi charged 50p per km. Set up a linear equation for this situation.

Solution

Distance travelled by bus: b km

Distance travelled by taxi: t km

Cost: $30b + 50t$ pence

or £$(0 \cdot 3b + 0 \cdot 5t)$

so $0 \cdot 3b + 0 \cdot 5t = 15$

Solving a problem graphically

Three burgers and two pizzas cost £12.

Using letters: 1 burger costs £x

1 pizza costs £y

Giving a total cost of £$(3x + 2y)$.

So $3x + 2y = 12$

This linear equation can be graphed as shown in the diagram opposite.

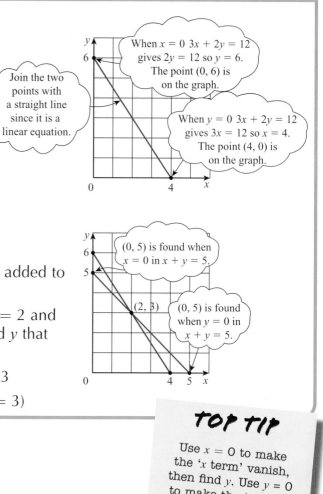

Join the two points with a straight line since it is a linear equation.

When $x = 0$ $3x + 2y = 12$ gives $2y = 12$ so $y = 6$. The point $(0, 6)$ is on the graph.

When $y = 0$ $3x + 2y = 12$ gives $3x = 12$ so $x = 4$. The point $(4, 0)$ is on the graph.

If you also know that one burger and one pizza together cost £5 then:

using letters: $x + y = 5$

The graph of this linear equation can be added to the diagram, as shown.

$(0, 5)$ is found when $x = 0$ in $x + y = 5$.

$(0, 5)$ is found when $y = 0$ in $x + y = 5$.

The two lines intersect at $(2, 3)$ giving $x = 2$ and $y = 3$ as the only pair of values for x and y that satisfy both equations simultaneously.

So a burger costs £2 and a pizza costs £3

$(x = 2)$ $(y = 3)$

TOP TIP

You should always explain clearly what x and y stand for.

TOP TIP

Use $x = 0$ to make the 'x term' vanish, then find y. Use $y = 0$ to make the 'y term' vanish then find x!

Quick Test 19

1. Set up a linear equation for each of these situations. Remember to state clearly the quantity represented by each letter you use.

   a) Three adults and four children go to the theatre. Their tickets cost £54 in total.

   b) The total cost of my journey was £18. The bus cost 25p per km and the taxi charged 65p per km.

2. Two families visit a theme park. Mr Clark and his son were charged £10 in total. Mr and Mrs Peters and their three children were charged a total of £24.

   a) Set up two algebraic equations to illustrate the cost for each family.

   b) By graphing these equations find the cost for 1 adult and for 1 child.

   c) How much would Mrs Walker and her five children be charged?

Working with simultaneous equations algebraically

Solving simultaneous equations algebraically

The general method is as follows:

Step 1 Rearrange the equations (if necessary) to get the letters lined up.

Step 2 Multiply the equations with the aim of matching the coefficient of one of the letters in each equation.

Step 3 Add or subtract to eliminate the matching letter then solve the resulting equation.

Step 4 Put the value you found in Step 3 back into one of the original equations to find the value of the other letter.

Step 5 Check that your solution satisfies both the equations.

Example
Solve $\left. \begin{array}{l} 2x = 7 - 3y \\ 3x - 2y - 4 \end{array} \right\}$

Solution

Rearrange the 1st equation

Step 1 $\left. \begin{array}{l} 2x + 3y = 7 \\ 3x - 2y = 4 \end{array} \right\}$ *x's and y's are now lined up*

Step 2 Let's match the x's:

$\left. \begin{array}{l} 2x + 3y = 7 \\ 3x - 2y = 4 \end{array} \right\} \begin{array}{l} \times 3 \\ \times 2 \end{array}$

This gives:

$\left. \begin{array}{l} 6x + 9y = 21 \\ 6x - 4y = 8 \end{array} \right\}$ *the x coefficients match in each equation*

Step 3 $\begin{array}{l} 6x + 9y = 21 \\ \underline{6x - 4y = 8} \end{array}$

Subtract: $13y = 13$ *subtract $-4y$ same as adding $4y$*

So $y = 1$

Step 4 Put $y = 1$ into $2x + 3y = 7$ giving

$2x + 3 = 7$

so $2x = 4$

so $x = 2$

The solution is $x = 2$ and $y = 1$.

Step 5   Check:

$2x = 7 - 3y$ becomes $2 \times 2 = 7 - 3 \times 1$ ✓

$3x - 2y = 4$ becomes $3 \times 2 - 2 \times 1 = 4$ ✓

An alternative method (substitution)

This method works in special cases when one of the equations gives one of the variables in terms of the other. For example for the equations

$y = 2x - 1$

$5x - 2y = 6$

you can replace y in $5x - 2y = 6$ by $2x - 1$

$5x - 2(2x - 1) = 6$

so $5x - 4x + 2 = 6$

so $x + 2 = 6$

this gives $x = 4$.

Now use $y = 2x - 1$

so $y = 2 \times 4 - 1 = 8 - 1 = 7$.

The solution is $x = 4$ and $y = 7$.

Example

Solve $\left. \begin{array}{l} a = 8 - 3b \\ 7a + 4b = 5 \end{array} \right\}$

Solution

Substituting $a = 8 - 3b$

in $7a + 4b = 5$ gives:

$7(8 - 3b) + 4b = 5$

so $56 - 21b + 4b = 5$

so $56 - 17b = 5$

so $51 = 17b$ giving $b = 3$

now using $a = 8 - 3b$

gives $a = 8 - 3 \times 3 = 8 - 9 = -1$.

The solution is $a = -1$ and $b = 3$.

> **TOP TIP**
>
> $3x + y = 8$ can be rearranged to $y = 8 - 3x$ and the substitution method used.

Solving a problem algebraically

Problem

A music station on the radio allows a fixed length of time for singles tracks and a longer fixed time for album tracks. One of the DJs knows that in his 35-minute programme he can fit 3 album tracks and 5 singles tracks. He also knows that the time allocated for 7 singles tracks is 18 minutes more than the time allocated for 2 album tracks.

Tomorrow he is broadcasting a half-hour programme and plans to play 3 album tracks and 4 singles tracks. Will he manage this?

Solution

The time allocated for 1 album track: a minutes.

The time allocated for 1 singles track: s minutes.

So $3a + 5s = 35$ (3 album tracks and 5 singles tracks in a 35-minute programme).

Also $7s = 2a + 18$ (7 singles tracks are 18 minutes longer than 2 album tracks).

This gives:

Step 1 $$5s + 3a = 35 \\ 7s - 2a = 18$$

TOP TIP

At Step 3 if the matching letters have: equal signs then **subtract** opposite signs then **add**.

Step 2 $$5s + 3a = 35 \quad \times 2 \rightarrow 10s + 6a = 70 \\ 7s - 2a = 18 \quad \times 3 \rightarrow 21s - 6a = 54$$

Step 3 $$10s + 6a = 70 \\ \underline{21s - 6a = 54}$$

Add: $31s \quad = 124$

So $s = 4$

Step 4 Put $s = 4$ in $5s + 3a = 35$

giving $20 + 3a = 35$ so $3a = 15$

so $a = 5$

album track: 5 minutes   singles track: 4 minutes

His plan needs $3a + 4s = 3 \times 5 + 4 \times 4 = 15 + 16 = 31$ minutes, 1 minute too long!

Step 5 Check: $3a + 5s = 35$ becomes $3 \times 5 + 5 \times 4 = 35$ ✓

$7s = 2a + 18$ becomes $7 \times 4 = 2 \times 5 + 18$ ✓

Quick Test 20

1. Using any suitable method, solve algebraically the system of equations:

   a) $2x + 3y = 13$
   $3x + 2y = 12$

   b) $y = 21 - 4x$
   $3x + 4y = 19$

   c) $5m + 3n = 15$
   $6m - 5n = 61$

2. Zoe and Mia buy fruit at the market.

   a) Zoe buys 2 apples and 3 oranges for £1·65. Write down an algebraic equation to illustrate this.

   b) Mia buys 5 apples and 4 oranges for £2·90. Write down an algebraic equation to illustrate this.

   c) By solving algebraically the system of equations you have written down, find the cost of 3 apples and 2 oranges.

Changing the subject of a formula

Why change the subject of a formula?

$C = \pi D$ The **subject** of this formula is C.

If you know the Diameter D then multiply by π to find the Circumference C.

$\dfrac{C}{\pi} = \dfrac{\pi D}{\pi}$ (dividing both sides by π) so $\dfrac{C}{\pi} = D$ (cancelling the factor π) giving $D = \dfrac{C}{\pi}$ (writing the formula the other way round).

$D = \dfrac{C}{\pi}$ The **subject** of this formula is D.

If you know the circumference C then divide by π to find the diameter D.

You change the subject depending on what you know and what you want to find!

Some basic examples

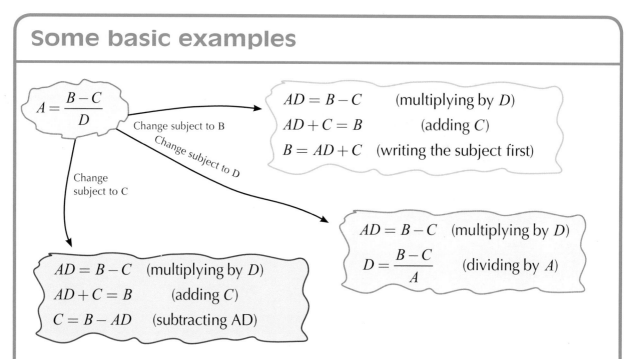

*All operations are performed on **both sides** of the equations.*

Example
The mean M of two numbers x and y is given by $M = \dfrac{x+y}{2}$.
Change the subject to x.

Solution
$2M = x + y$ (multiplying by 2)
$2M - y = x$ (subtracting y)
$x = 2M - y$ (writing the subject first)

Working with brackets

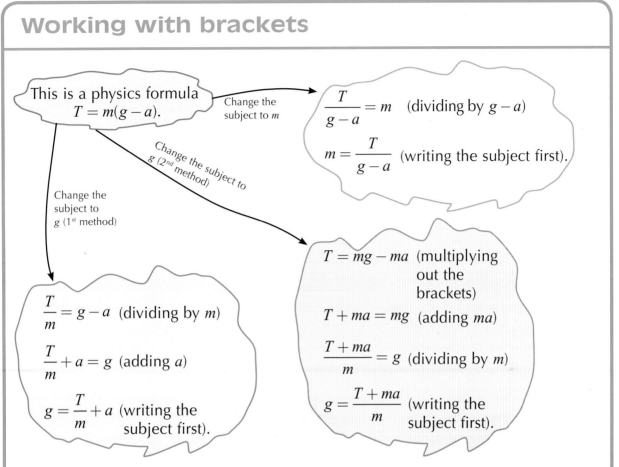

This is a physics formula
$T = m(g - a)$.

Change the subject to m

$\dfrac{T}{g-a} = m$ (dividing by $g - a$)

$m = \dfrac{T}{g-a}$ (writing the subject first).

Change the subject to g (2nd method)

Change the subject to g (1st method)

$\dfrac{T}{m} = g - a$ (dividing by m)

$\dfrac{T}{m} + a = g$ (adding a)

$g = \dfrac{T}{m} + a$ (writing the subject first).

$T = mg - ma$ (multiplying out the brackets)

$T + ma = mg$ (adding ma)

$\dfrac{T + ma}{m} = g$ (dividing by m)

$g = \dfrac{T + ma}{m}$ (writing the subject first).

*All operations are performed on **both sides** of the equations.*

Note:

$$\dfrac{T + ma}{m}$$

$$= \dfrac{T}{m} + \dfrac{ma}{m} \text{ (divide both terms by } m\text{)}$$

$$= \dfrac{T}{m} + a \text{ (cancel the factor } m\text{)}.$$

Check using simple numbers: $\dfrac{1 + 3}{4} = \dfrac{1}{4} + \dfrac{3}{4} = 1$

Working with squares and square roots

Let's take the formula $v = \sqrt{9 - u^2}$ and change the subject to u.

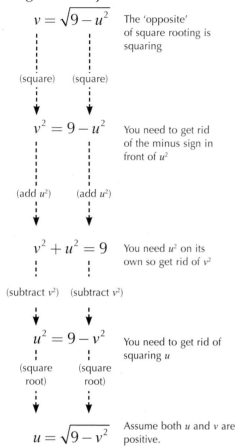

$v = \sqrt{9 - u^2}$ The 'opposite' of square rooting is squaring

(square)   (square)

$v^2 = 9 - u^2$ You need to get rid of the minus sign in front of u^2

(add u^2)   (add u^2)

$v^2 + u^2 = 9$ You need u^2 on its own so get rid of v^2

(subtract v^2)   (subtract v^2)

$u^2 = 9 - v^2$ You need to get rid of squaring u

(square root)   (square root)

$u = \sqrt{9 - v^2}$ Assume both u and v are positive.

Example

Change the subject of $r = \sqrt{\dfrac{A}{\pi}}$ to A.

Solution

$r^2 = \dfrac{A}{\pi}$ after squaring both sides

$\pi r^2 = A$ after multiplying both sides by π

$A = \pi r^2$ writing the subject first

Do you recognise this formula?

Example

Change the subject of $E = \dfrac{1}{2}mv^2$ to v.

Solution

$2E = mv^2$ after doubling each side

$\dfrac{2E}{m} = v^2$ after dividing by m

$v = \sqrt{\dfrac{2E}{m}}$ after square rooting both sides and writing the subject first. Assume v is positive.

u is now the subject.

Note: Square rooting both sides should introduce the possibility of a negative. For example:

$x^2 = 9 \xrightarrow[\text{both sides}]{\text{square root}} x = \pm 3$ (since $3^2 = 9$ and also $(-3)^2 = -3 \times (-3) = 9$)

Quick Test 21

1. In each case change the subject of the formula to the letter indicated:

   a) $M = \dfrac{A}{s}$ to s

   b) $P = \dfrac{x + y}{c}$ to y

   c) $a^2 + b^2 = c^2$ to a

2. By multiplying out the brackets in the formula $y - b = m(x - a)$ change the subject to x.

3. The formula $F = \dfrac{9}{5}C + 32$ is useful for changing °C to °F. Work out a formula that is useful for changing °F to °C.

Quadratic equations

What is a quadratic equation?

Examples

$2x^2 - 3x + 4 = 0$

$5 - 3x^2 = 0 \qquad x^2 = 4$

$\qquad 5x^2 = x$

These are equations that have an 'x^2 term'.

They might have an 'x-term' or maybe not.

They may have a 'constant term' or maybe not.

There should be no other types of terms.

All quadratic equations can be arranged into this 'standard form':

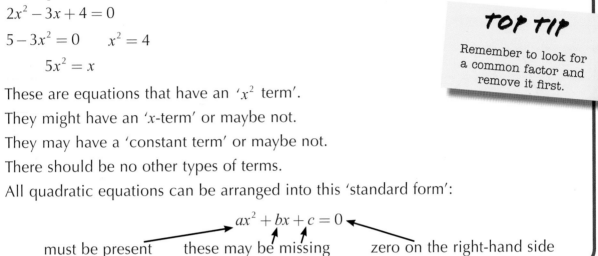

$$ax^2 + bx + c = 0$$

must be present   these may be missing   zero on the right-hand side

> **TOP TIP**
>
> Remember to look for a common factor and remove it first.

Solving quadratic equations by factorising

If the equation is not in 'standard form' $ax^2 + bx + c = 0$ then you should rearrange it. You need a zero on the right-hand side of the equation for this method!

The general method is:

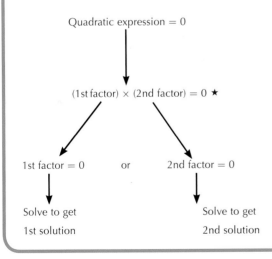

Quadratic expression = 0

(1st factor) × (2nd factor) = 0 ★

1st factor = 0   or   2nd factor = 0

Solve to get   Solve to get

1st solution   2nd solution

Note 1

At step ★ you use the fact that if you multiply two numbers to get zero then one or other of the two numbers is zero:

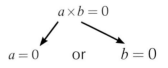

$$a \times b = 0$$

$a = 0$ or $b = 0$

Note 2

A solution of an equation is called a **root** of the equation.

Example 1

Solve $x^2 - 2x - 3 = 0$

Solution

Factorise the quadratic expression $x^2 - 2x - 3$: see page 26.

$$x^2 - 2x - 3 = 0$$
$$(x + 1)(x - 3) = 0$$

$x + 1 = 0$ or $x - 3 = 0$
$\quad x = -1 \qquad\qquad x = 3$

The two roots of the equation are −1 and 3.

Example 2

Find the roots of the following quadratic equation:

$$3x^2 + 5x - 2 = 0$$

Solution

$$3x^2 + 5x - 2 = 0$$
$$(3x - 1)(x + 2) = 0$$

$3x - 1 = 0$ or $x + 2 = 0$
$3x = 1 \qquad\qquad x = -2$
$\quad x = \frac{1}{3}$

The roots are $\frac{1}{3}$ and −2.

Solving quadratic equations by formula

When solving $ax^2 + bx + c = 0$, sometimes $(\ ?\ )(\ ?\ ) = 0$ (the factorising step) does not work, i.e. the expression does not factorise.

In this case you use the **quadratic formula**:

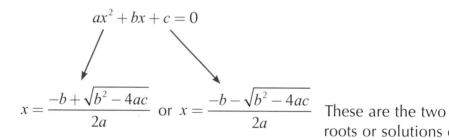

$$ax^2 + bx + c = 0$$

$$x = \frac{-b + \sqrt{b^2 - 4ac}}{2a} \quad \text{or} \quad x = \frac{-b - \sqrt{b^2 - 4ac}}{2a}$$

These are the two roots or solutions of the equation.

The compact way of writing this is $x = \dfrac{-b \pm \sqrt{b^2 - 4ac}}{2a}$ ± means there are two possibilities: one from adding, the other from subtracting.

Example

Solve $3x^2 - 4x - 2 = 0$ using an appropriate formula giving your answer correct to 1 decimal place.

Solution

$$3x^2 - 4x - 2 = 0$$

compare $ax^2 + bx + c = 0$

This gives $\quad a = 3$ (the number of 'x^2's)

$\qquad b = -4$ (the number of 'x's)

and $\quad c = -2$ (the constant term)

> **TOP TIP**
>
> If the question asks you to solve a quadratic equation and mentions rounding, e.g. to 1 decimal place, then you know to use the formula!

The appropriate formula is $x = \dfrac{-b \pm \sqrt{b^2 - 4ac}}{2a}$

which becomes: $x = \dfrac{-(-4) \pm \sqrt{(-4)^2 - 4 \times 3 \times (-2)}}{2 \times 3} = \dfrac{4 \pm \sqrt{16 + 24}}{6}$

$$= \dfrac{4 \pm \sqrt{40}}{6}$$

so $x = \dfrac{4 + \sqrt{40}}{6} \doteqdot 1 \cdot 7$ or $x = \dfrac{4 - \sqrt{40}}{6} \doteqdot -0 \cdot 4$

$\qquad\qquad$ (to 1 d.p.) $\qquad\qquad\qquad$ (to 1 d.p.)

Quick Test 22

1. Solve these equations algebraically:

    a) $x^2 + 2x - 15 = 0$ $\quad$ b) $15x^2 + 11x + 2 = 0$

2. Solve these equations algebraically:

    a) $8x - 2x^2 = 0$ $\qquad$ b) $x^2 = 6x$

3. Solve these equations, giving your answer correct to 1 decimal place:

    a) $x^2 + 4x + 1 = 0$ $\qquad$ b) $4x^2 - 12x + 3 = 0$

Quadratic graphs

The parabola

A graph showing the values of x^2 for all values of x can be built up from a few particular values:

x	-3	-2	-1	0	1	2	3
$y = x^2$	9	4	1	0	1	4	9

(−3, 9) (−1, 1) (1, 1) (3,9)

(−2, 4) (0, 0) (2,4)

Notes

(1) This type of graph shape is called a **parabola**.

(2) The graph has symmetry, with the y-axis (the line $x = 0$) being the axis of symmetry.

(3) The graph has a minimum turning point at the origin (0, 0). This means that x^2 has a minimum value of 0 when $x = 0$.

(4) The equation $y = 0$ or $x^2 = 0$ has one **solution (root)**, namely $x = 0$.

The graph $y = x^2$

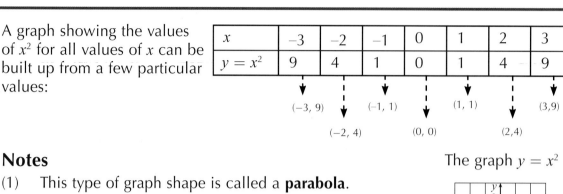

The graph $y = kx^2$

For $k > 0$ (positive) the graph is concave upwards:

Minimum point (0, 0)

For $k < 0$ (negative) the graph is concave downwards:

Maximum point (0, 0)

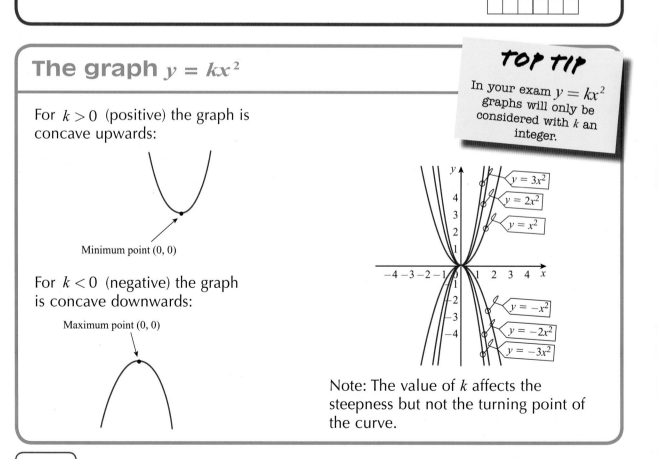

TOP TIP

In your exam $y = kx^2$ graphs will only be considered with k an integer.

Note: The value of k affects the steepness but not the turning point of the curve.

Example

Use the information in the diagram to calculate the value of k.

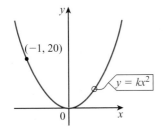

Solution

$(-1, 20)$ lies on the curve so $x = -1$ and $y = 20$ satisfy the equation

$y = kx^2$

This gives: $20 = k \times (-1)^2$

so $20 = k \times 1$

and therefore $k = 20$.

The graphs $y = (x + a)^2 + b$ and $y = (x - a)^2 + b$

The graph $y = x^2$ (see previous page) is moved to get:

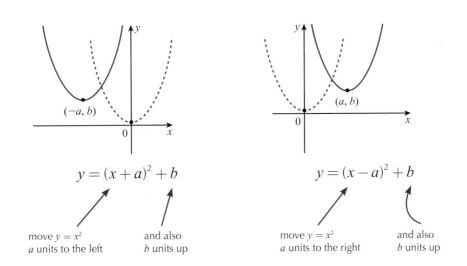

$y = (x + a)^2 + b$

move $y = x^2$ and also
a units to the left b units up

$y = (x - a)^2 + b$

move $y = x^2$ and also
a units to the right b units up

Note: Given a quadratic expression like $x^2 + 4x + 7$ you can 'complete the square' (see page 28) to get $(x + 2)^2 + 3$ and 'read off' the minimum turning point $(-2, 3)$.

Example

Give the coordinates of the minimum turning point of the graph $y = x^2 - 6x + 10$.

Solution

$y = x^2 - 6x + 10 = (x - 3)^2 + 1$ so $y = x^2$ is moved 3 units to the right and 1 unit up:

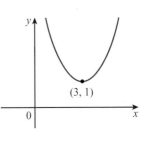

The minimum turning point is $(3, 1)$.

Quadratic graphs and factors

To find the x-axis intercepts you set $y = 0$. For example the graph $y = 2x^2 - x - 3$ crosses the x-axis when $y = 2x^2 - x - 3 = 0$.

Solving this quadratic equation gives:

$(2x - 3)(x + 1) = 0$ so $x = \frac{3}{2}$ and $x = -1$ (see page 67).

The 'x^2-term' is positive $(2x^2)$ and so the graph is 'concave upwards' (see middle of page 70).

The y-axis intercept is found when you set $x = 0$ in the equation.

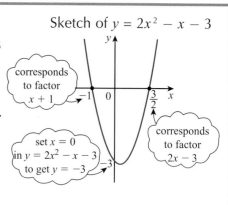

Sketch of $y = 2x^2 - x - 3$

corresponds to factor $x + 1$

corresponds to factor $2x - 3$

set $x = 0$ in $y = 2x^2 - x - 3$ to get $y = -3$

Sketching quadratic graphs – hints

- $y = ax^2 + bx + c$ If $a > 0$ (positive) the graph is 'concave upwards'.
  If $a < 0$ (negative) the graph is 'concave downwards'.
- Where does it cross the y-axis? → set $x = 0$ to find the value of y.

  Where does it cross the x-axis? → set $y = 0$ and solve the resulting equation.
- Complete the square to get $y = (x \pm a)^2 + b$ (see page 71) to find the turning point.
- Plot a few points: choose a value for x and calculate y.

TOP TIP

Calculating the **discriminant** $b^2 - 4ac$ tells you a lot about the graph – see page 74.

Quick Test 23

1. Use the information in the diagram to calculate the value of k.

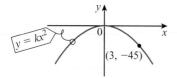

$y = kx^2$

$(3, -45)$

2. Give the coordinates of the minimum turning point on the graph $y = x^2 + 10x + 28$.

3. Sketch the graph of $y = 3x^2 - 10x - 8$ showing clearly the x-axis and y-axis intercepts.

Working with quadratic functions

Maximum and minimum values

From a quadratic graph that crosses the x-axis, you can find the maximum or minimum value on the graph as follows:

for a negative x^2-term the graph is concave downwards.

Note: The dotted line is *an axis* of symmetry of the parabola.

Use the 'halfway' x value in the graph equation to calculate the y-coordinate for max/min value.

In a similar way you can find the minimum value on a concave upwards graph (one with a positive x^2-term).

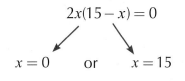

The y-coordinate is the maximum value.

$$x = \frac{a+b}{2}$$

(halfway between a and b)

Example

It is known that the cross-sectional area, A cm^2, of this guttering is given by $A(x) = 30x - 2x^2$.

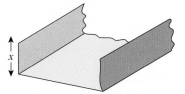

Find the maximum value of this area.

Here is a sketch of the graph.

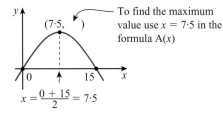

To find the maximum value use $x = 7 \cdot 5$ in the formula $A(x)$

$$x = \frac{0+15}{2} = 7 \cdot 5$$

Solution

The graph showing the values of the area will be a parabola. To find where it crosses the x-axis, you solve $A(x) = 0$.

so
$$30x - 2x^2 = 0$$
$$2x(15 - x) = 0$$

$$x = 0 \quad \text{or} \quad x = 15$$

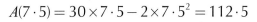

TOP TIP

The equation of a line parallel to the y-axis has the form $x = k$ where k is a constant.

$$A(7 \cdot 5) = 30 \times 7 \cdot 5 - 2 \times 7 \cdot 5^2 = 112 \cdot 5$$

The maximum cross-sectional area is **112·5 cm^2**.

The discriminant

Solving $ax^2 + bx + c = 0$ using the quadratic formula (see page 68) apparently gives two roots (solutions):

$$x = \frac{-b + \sqrt{b^2 - 4ac}}{2a} \text{ and } x = \frac{-b - \sqrt{b^2 - 4ac}}{2a}$$

However there may not be two! It depends on the value $b^2 - 4ac$ under the square root sign, $b^2 - 4ac$ is called the **discriminant**. If it's zero then '$+\sqrt{0}$' and '$-\sqrt{0}$' are the same so there's only one root. If it's negative then $\sqrt{\text{negative number}}$ does not give a Real number value.

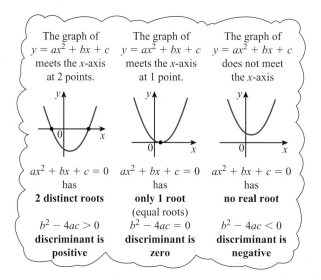

The graph of $y = ax^2 + bx + c$ meets the x-axis at 2 points.	The graph of $y = ax^2 + bx + c$ meets the x-axis at 1 point.	The graph of $y = ax^2 + bx + c$ does not meet the x-axis
$ax^2 + bx + c = 0$ has **2 distinct roots**	$ax^2 + bx + c = 0$ has **only 1 root** (equal roots)	$ax^2 + bx + c = 0$ has **no real root**
$b^2 - 4ac > 0$ **discriminant is positive**	$b^2 - 4ac = 0$ **discriminant is zero**	$b^2 - 4ac < 0$ **discriminant is negative**

Note: If x^2 term is negative the graphs will be *concave* downwards.

Example
Find the range of values of p such that $2x^2 + x - p = 0$ has no Real roots.

Solution

$$\text{Compare } 2x^2 + x - p = 0$$
$$\text{with } ax^2 + bx + c = 0$$
$$\text{so } a = 2, b = 1, c = -p.$$

$\boxed{\text{Discriminant} \atop (b^2 - 4ac)} = 1^2 - 4 \times 2 \times (-p) = 1 + 8p$

For no Real roots: $1 + 8p < 0$ so $8p < -1$ giving $p < -\frac{1}{8}$.

Turning points and the axis of symmetry

The axis of symmetry of a parabola passes through its turning point.

For example:

$y = 3 - (x + 2)^2$.

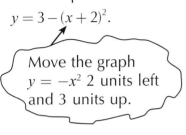
Move the graph $y = -x^2$ 2 units left and 3 units up.

The x^2-term is negative so the graph is concave downwards.
It has a maximum turning point at $(-2, 3)$.
The axis of symmetry is $x = -2$.

Sketch of $y = 3 - (x + 2)^2$.

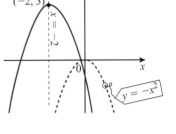

Graphic calculators

Graphic calculators are great for checking results. Usually a mistake in your calculation can be detected from a quick graph.

For example, if you have just solved $3x^2 - 4x - 2 = 0$ using the quadratic formula then your solutions are $x = 1 \cdot 72$ and $x = -0 \cdot 39$ (to 2 decimal places).

Warning

Results written down directly from your graphic calculator will gain no marks. Full working is required. A graphic calculator is only useful for detecting possible errors in your calculation of the solution to an equation.

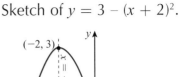

Calculator tip

Graph: $y = 3x^2 - 4x - 2$
with   Xmin: −1   Ymin: −5
       max: 2   max: 5
       scl: 1   scl: 1
These intersections seem reasonable for $x \doteqdot -0 \cdot 4$ and $x \doteqdot 1 \cdot 7$. You could 'zoom in' to check the results of your calculation more accurately.

TOP TIP

when calculating $b^2 - 4ac$ be careful with negative values.

Quick Test 24

1. $f(x) = 12 + x - x^2$ is a quadratic function. By solving $f(x) = 0$ find the value of x for which f has a maximum value and hence find that maximum value.

2. Find the range of values of k for which the equation $kx^2 - 2x - 1 = 0$ has two distinct Real roots.

3. For the graph $y = 1 - (x - 2)^2$ find:

   a) The coordinates of the maximum turning point.

   b) The equation of the axis of symmetry.

Pythagoras' Theorem and its converse

Pythagoras' Theorem

TOP TIP

The side opposite the right angle is the longest side and is called the **hypotenuse**.

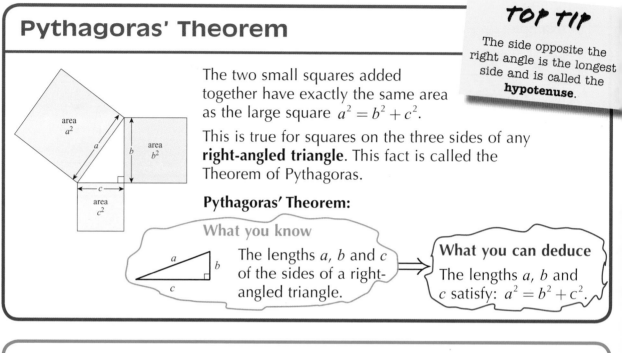

The two small squares added together have exactly the same area as the large square $a^2 = b^2 + c^2$.

This is true for squares on the three sides of any **right-angled triangle**. This fact is called the Theorem of Pythagoras.

Pythagoras' Theorem:

What you know

The lengths a, b and c of the sides of a right-angled triangle.

What you can deduce

The lengths a, b and c satisfy: $a^2 = b^2 + c^2$.

Calculating side lengths

You can use Pythagoras' Theorem to calculate the lengths of the sides of a right-angled triangle:

$$a^2 = b^2 + c^2 \quad \text{or} \quad b^2 = a^2 - c^2 \quad \text{or} \quad c^2 = a^2 - b^2$$

$$a = \sqrt{b^2 + c^2} \quad \text{or} \quad b = \sqrt{a^2 - c^2} \quad \text{or} \quad c = \sqrt{a^2 - b^2}$$

Length of the largest side

Length of one of the smaller sides

Remember

Add squares then $\boxed{\sqrt{}}$: hypotenuse

Subtract squares then $\boxed{\sqrt{}}$: smaller side

Example Calculate BC to 1 d.p.

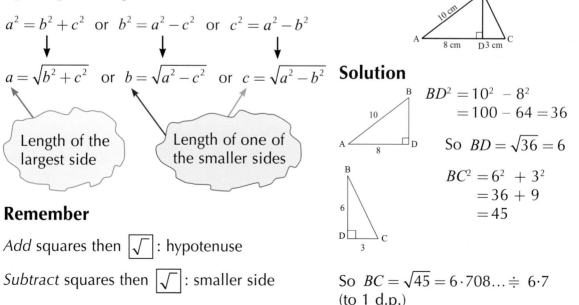

Solution

$$BD^2 = 10^2 - 8^2$$
$$= 100 - 64 = 36$$

So $BD = \sqrt{36} = 6$

$$BC^2 = 6^2 + 3^2$$
$$= 36 + 9$$
$$= 45$$

So $BC = \sqrt{45} = 6 \cdot 708... \doteqdot 6 \cdot 7$
(to 1 d.p.)

Calculations in context

Example

A house has a basic shape of a cuboid surmounted by a triangular prism.

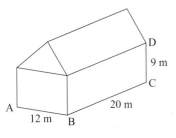

How long is a string stretching from A to D inside the house?

Solution

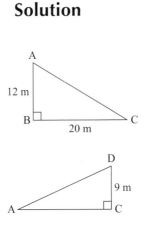

First calculate AC using $\triangle$ABC (on the floor of the house!)

$$AC^2 = 12^2 + 20^2 = 144 + 400$$
$$= 544$$

(don't take the square root yet)

$$AD^2 = AC^2 + 9^2$$
$$= 544 + 81 = 625$$

so $\quad AD = \sqrt{625} = 25$ m

Note: Here are the triangles shown by dotted lines:

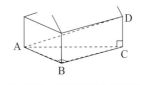

An application of Pythagoras' Theorem

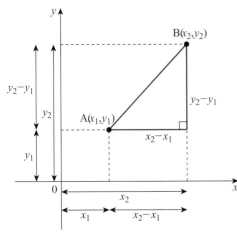

Using Pythagoras' Theorem:

$$AB^2 = (x_2 - x_1)^2 + (y_2 - y_1)^2$$

so $\quad AB = \sqrt{(x_2 - x_1)^2 + (y_2 - y_1)^2}$

↑ x-coordinate difference squared

↑ y-coordinate difference squared

Note: You are not required to learn this formula but you should understand how Pythagoras' Theorem is used to work it out!

Example

Find the distance AB where A(–3, 1) and B(3, 9).

Solution

$$AB = \sqrt{(3 - (-3))^2 + (9 - 1)^2}$$
$$= \sqrt{6^2 + 8^2}$$
$$= \sqrt{36 + 64} = \sqrt{100} = 10$$

so AB = 10 units

The converse of Pythagoras' Theorem

The converse of Pythagoras' Theorem:

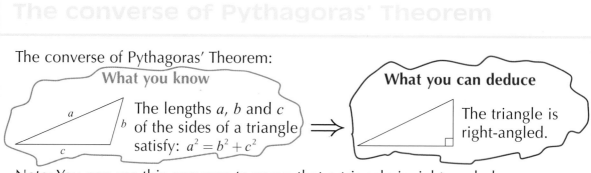

What you know

The lengths a, b and c of the sides of a triangle satisfy: $a^2 = b^2 + c^2$ $\Longrightarrow$

What you can deduce

The triangle is right-angled.

Note: You can use this converse to prove that a triangle is right-angled.

Example

A $18\cdot6\,\text{cm} \times 24\cdot8\,\text{cm}$ picture was framed. The diagonal was measured at 31cm.

Is the frame rectangular?

Solution

$AB^2 + BC^2 = 18\cdot6^2 + 24\cdot8^2 = 961$

also $AC^2 = 31^2 = 961$

so $AB^2 + BC^2 = AC^2$

and using the converse of Pythagoras' Theorem triangle ABC is right-angled at B. The frame is rectangular.

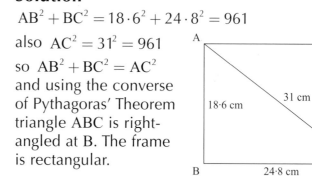

TOP TIP

When you prove a triangle is right-angled state clearly you are using the converse of Pythagoras' Theorem.

Quick Test 25

1. Calculate x correct to 1 decimal place:

   a)

   2·3 cm

   1·9 cm x

   b)

   10·2 cm   6·1 cm

   x

2. Is this a right-angled triangle?

   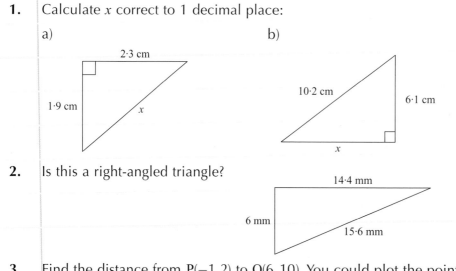

   14·4 mm

   6 mm

   15·6 mm

3. Find the distance from P(-1, 2) to Q(6, 10). You could plot the points as a diagram, complete a right-angle triangle and use Pythagoras' Theorem.

4. Calculate (to 3 s.f.) the length of a space diagonal (from one vertex through the centre to the opposite vertex) of a $3\,\text{cm} \times 3\,\text{cm} \times 3\,\text{cm}$ cube.

Working with angles and polygons

Some basic angle facts

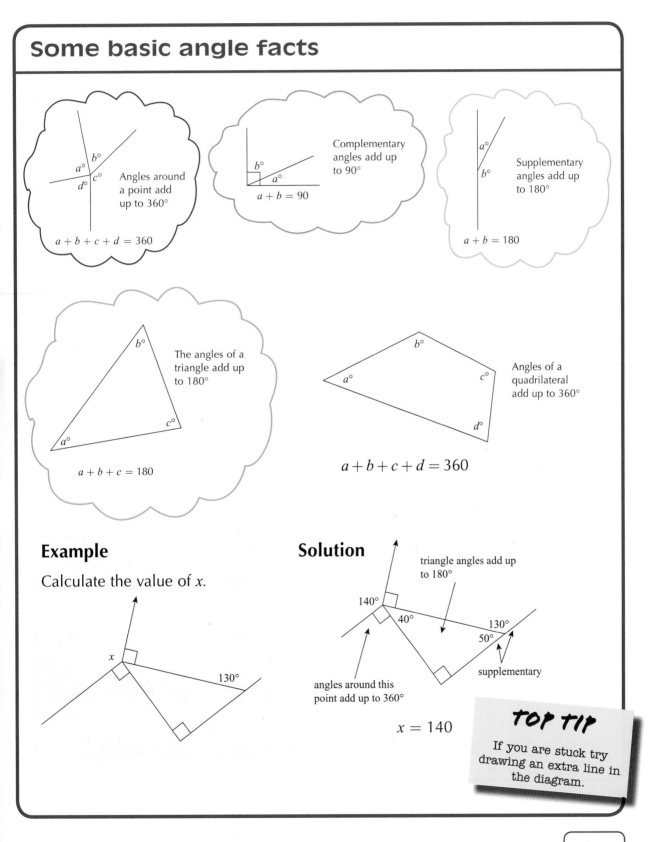

Angles around a point add up to 360°

$a + b + c + d = 360$

Complementary angles add up to 90°

$a + b = 90$

Supplementary angles add up to 180°

$a + b = 180$

The angles of a triangle add up to 180°

$a + b + c = 180$

Angles of a quadrilateral add up to 360°

$a + b + c + d = 360$

Example

Calculate the value of x.

Solution

triangle angles add up to 180°

angles around this point add up to 360°

supplementary

$x = 140$

TOP TIP

If you are stuck try drawing an extra line in the diagram.

Angles and parallel lines

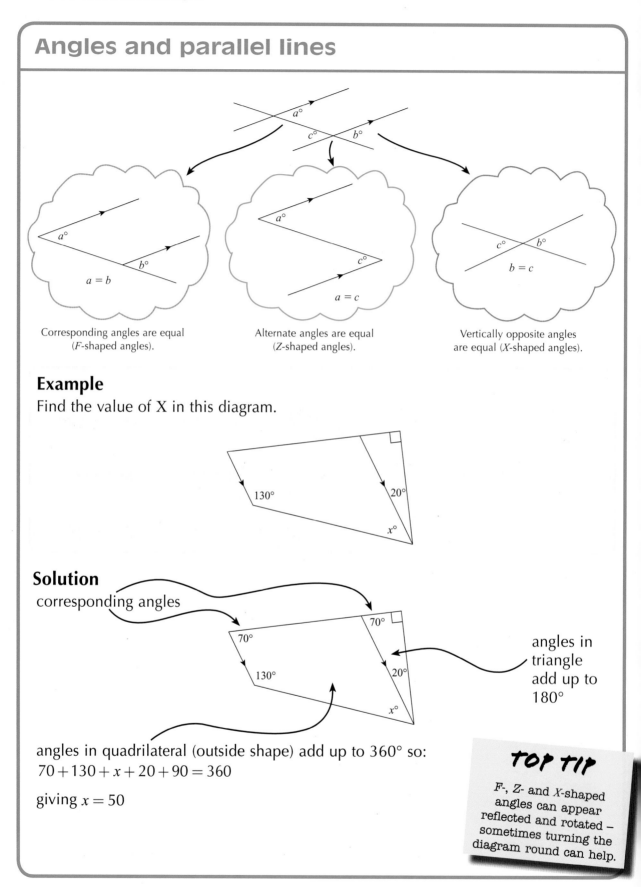

Corresponding angles are equal
(*F*-shaped angles).

$a = b$

Alternate angles are equal
(*Z*-shaped angles).

$a = c$

Vertically opposite angles
are equal (*X*-shaped angles).

$b = c$

Example

Find the value of X in this diagram.

Solution

corresponding angles

angles in triangle add up to 180°

angles in quadrilateral (outside shape) add up to 360° so:

$$70 + 130 + x + 20 + 90 = 360$$

giving $x = 50$

TOP TIP

F-, *Z*- and *X*-shaped angles can appear reflected and rotated – sometimes turning the diagram round can help.

Types of triangles

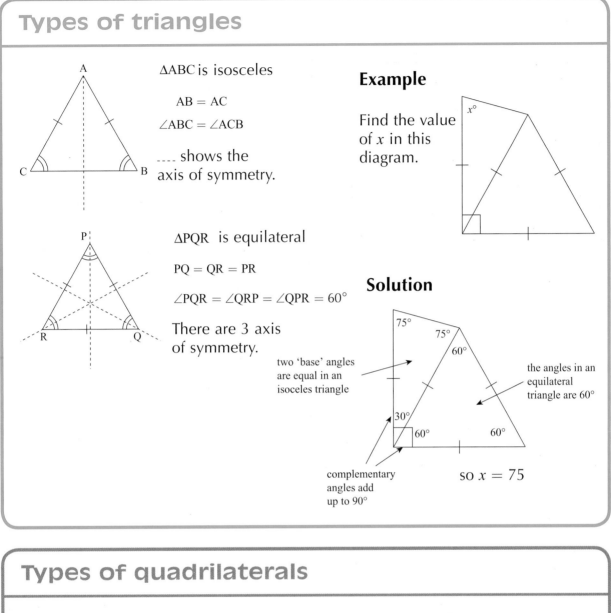

△ABC is isosceles

AB = AC

∠ABC = ∠ACB

.... shows the axis of symmetry.

△PQR is equilateral

PQ = QR = PR

∠PQR = ∠QRP = ∠QPR = 60°

There are 3 axis of symmetry.

Example

Find the value of x in this diagram.

Solution

two 'base' angles are equal in an isoceles triangle

the angles in an equilateral triangle are 60°

complementary angles add up to 90°

so $x = 75$

Types of quadrilaterals

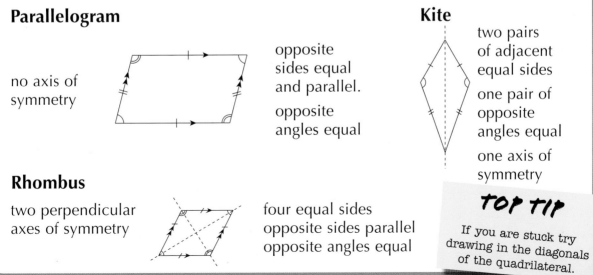

Parallelogram

no axis of symmetry

opposite sides equal and parallel.

opposite angles equal

Kite

two pairs of adjacent equal sides

one pair of opposite angles equal

one axis of symmetry

Rhombus

two perpendicular axes of symmetry

four equal sides
opposite sides parallel
opposite angles equal

TOP TIP

If you are stuck try drawing in the diagonals of the quadrilateral.

Regular polygons

A polygon is a shape with straight line edges.

Regular polygons have equal sides and equal angles.

No. of sides	3	4	5
Name	Equilateral triangle	Square	Pentagon

No. of sides	6	7	8	9	10
Name	Hexagon	Septagon	Octagon	Nonagon	Decagon

Example

Find the size of the inside angles of a regular pentagon.

Solution

You are trying to find $\angle DAB$.

The pentagon can be divided into 5 identical isosceles triangles, as shown.

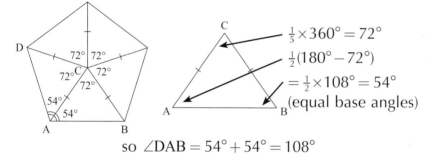

$\frac{1}{5} \times 360° = 72°$

$\frac{1}{2}(180° - 72°)$

$= \frac{1}{2} \times 108° = 54°$

(equal base angles)

so $\angle DAB = 54° + 54° = 108°$

Quick Test 26

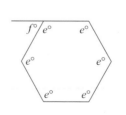

TOP TIP

If the same letter is used for the size of two different angles then they are equal.

1. Calculate the value of the letters in each diagram:

   a)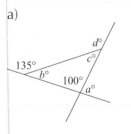

   b)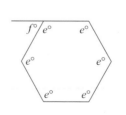

2.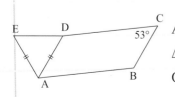

   ABCD is a parallelogram.

   ΔADE is isosceles.

   Calculate the size of $\angle EAB$.

Working with angles and circles

Tangents to circles

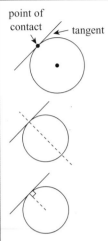

point of contact
tangent

A tangent is a line which touches a circle at only one point. This point is called the **point of contact**.

The line through the centre and the point of contact is an axis of symmetry for the diagram.

This means that a tangent and the radius to the point of contact are at right angles (perpendicular).

Two tangents can be drawn to a circle from a point outside the circle.

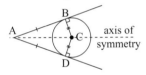

A · · · B · · · C · · · axis of symmetry · · · D

In the diagram, tangents AB and AD and the two radii CB and CD form a kite, ABCD.

Example

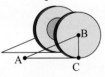

A cylindrical drum of radius 2 m is held in place by triangular metal supports on each side, as shown. Rod AB is 4 m long and attached to the centre of the circle B, directly above C. How many triangular metal frames like ABC can be made from a 100 m length of rod?

Solution

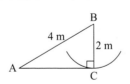

B
4 m
2 m
A
C

$BC = 2$ m (radius)
$\angle BCA = 90°$ (angle between tangent AC and radius BC)

So $AC^2 = 4^2 - 2^2 = 16 - 4 = 12$

giving $AC = \sqrt{12} = 3 \cdot 46\ldots$

Length of rod in one frame $= 4 + 2 + 3 \cdot 46\ldots$

$= 9 \cdot 46\ldots$

Total no. of frames $= \frac{100}{9 \cdot 46}\cdots = 10 \cdot 56\ldots$

10 frames only

(Don't round up since 11 frames would require over 100 m of rod.)

Angles in a semicircle

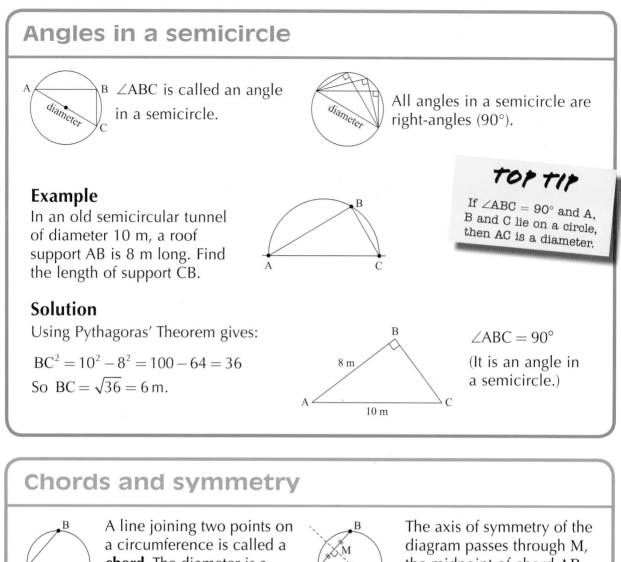

∠ABC is called an angle in a semicircle.

All angles in a semicircle are right-angles (90°).

TOP TIP

If ∠ABC = 90° and A, B and C lie on a circle, then AC is a diameter.

Example

In an old semicircular tunnel of diameter 10 m, a roof support AB is 8 m long. Find the length of support CB.

Solution

Using Pythagoras' Theorem gives:

$BC^2 = 10^2 - 8^2 = 100 - 64 = 36$

So $BC = \sqrt{36} = 6$ m.

∠ABC = 90°

(It is an angle in a semicircle.)

Chords and symmetry

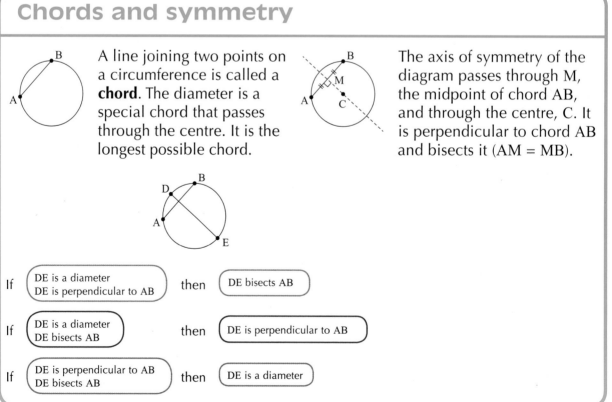

A line joining two points on a circumference is called a **chord**. The diameter is a special chord that passes through the centre. It is the longest possible chord.

The axis of symmetry of the diagram passes through M, the midpoint of chord AB, and through the centre, C. It is perpendicular to chord AB and bisects it (AM = MB).

If (DE is a diameter / DE is perpendicular to AB) then (DE bisects AB)

If (DE is a diameter / DE bisects AB) then (DE is perpendicular to AB)

If (DE is perpendicular to AB / DE bisects AB) then (DE is a diameter)

Example

A circular table of diameter 1·4 m is hinged to the wall along AB as shown. The table, when up, extends 1·3 m out from the wall. What length of hinge is required?

Solution

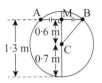

Diameter = 1·4 m so radius $= \dfrac{1\cdot4}{2} = 0\cdot7$ m.

Draw in the diameter perpendicular to chord AB. M is the midpoint of AB.

$$MC = 1\cdot3 - 0\cdot7 = 0\cdot6 \text{ m}$$
$$\text{(radius)}$$
$$CB = 0\cdot7 \text{ m (radius)}$$

Draw a radius CB to make a right-angled triangle.

TOP TIP

To bisect is to cut into two equal parts.

Use Pythagoras' Theorem:

$$MB^2 = 0\cdot7^2 - 0\cdot6^2 = 0\cdot13 \qquad \text{So } MB = \sqrt{0\cdot13} = 0\cdot360...$$

giving $AB = 2 \times MB$

$= 2 \times 0\cdot360...$

$= 0\cdot721...$

The hinge is **72 cm** (to the nearest cm).

Quick Test 27

TOP TIP

Pay attention to the accuracy of your answer or you may lose marks.

1. Calculate x to 3 significant figures:

   a)

   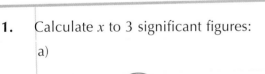

   AB is a tangent to a circle centre C.

   b)

   AB is a diameter and C lies on the circumference.

2. This cylindrical water tank has a radius of 1·9 metres and the maximum depth of the water is 3·1 metres. Find x, the width of the water surface, to the nearest centimetre.

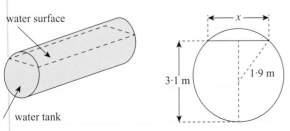

Similar shapes

What are similar shapes?

Shapes and solids may be enlarged or reduced in size. When this is done with all measurements kept in proportion, the resulting shape or solid is said to be mathematically **similar** to the original.

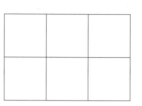

These rectangles are **SIMILAR**.

What is the scale factor?

The number given by $\frac{\text{new length}}{\text{old length}}$ is called the scale factor and is usually denoted by k.

For example, here is a triangle enlarged ($k > 1$):

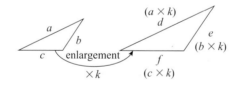

The scale factor $k = \frac{d}{a} = \frac{e}{b} = \frac{f}{c}$ ⟵ larger ⟵ smaller

For reductions: $0 < k < 1$
(the scale factor, k, lies between 0 and 1).

For enlargements: $k > 1$
(the scale factor, k, is greater than 1).

Here is the triangle reduced ($0 < k < 1$).

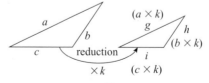

The scale factor $k = \frac{g}{a} = \frac{h}{b} = \frac{i}{c}$ ⟵ smaller ⟵ larger

Solving problems using similarity

TOP TIP

For triangles, if corresponding angles are equal (i.e. if the triangles are equiangular), then the triangles are similar.

Similar shapes have corresponding angles equal and sides in the same proportion.

Example

Emma, who is 1·9 m tall, stands $10\frac{1}{2}$ m from the base of a tree. She notices that her 3-metre-long shadow reaches exactly as far as the tree's shadow reaches. How tall is the tree?

Solution

Here is the diagram:

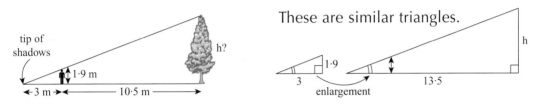

These are similar triangles.

Scale factor $= \frac{13\cdot5}{3} = 4\cdot5$ (enlargement scale factor greater than 1).

This means that any length in the large triangle is $4\cdot5$ times the corresponding length in the small triangle.

So $h = 1\cdot9 \times 4\cdot5 = 8\cdot55$.

The tree is 8 m 55 cm high.

Scale factors for area and volume

There are different scale factors at work in this enlargement: length, area and volume scale factors.

	measurement on small cube	measurement on large cube	effect
length of edge	1 unit	3 units	x 3
area of face	1 unit²	9 unit²	x 9
volume	1 unit³	27 unit³	x 27

enlargement

Scale factor for length is 3.

Scale factor for area is $3^2 (= 9)$.

Scale factor for volume is $3^3 (= 27)$.

In general:

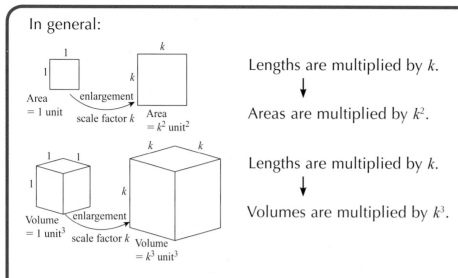

Lengths are multiplied by k.

Areas are multiplied by k^2.

Lengths are multiplied by k.

Volumes are multiplied by k^3.

Example

These two bottles are mathematically similar. The volume of the smaller bottle is 80 cl. Find the volume of the larger bottle in litres.

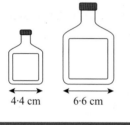

4·4 cm   6·6 cm

Solution

The (length) scale factor $= \frac{6 \cdot 6}{4 \cdot 4} = 1 \cdot 5$.

So the volume scale factor $= 1 \cdot 5^3$.

Small volume $= 80\ \text{cl} = 0 \cdot 8$ litres (centilitres).

Larger volume $= 0 \cdot 8 \times 1 \cdot 5^3 = 2 \cdot 7$ litres.

Quick Test 28

TOP TIP

Always separate the two similar triangles and redraw them, filling in all the lengths you know on the new diagrams.

1. a)

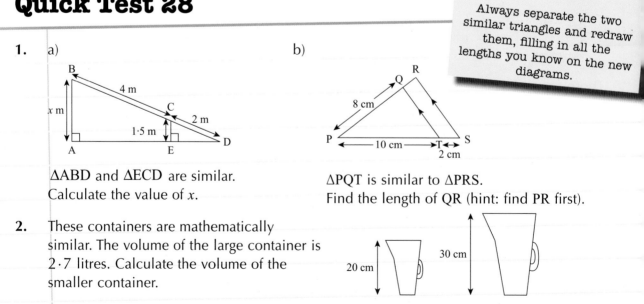

b)

$\triangle$ABD and $\triangle$ECD are similar.
Calculate the value of x.

$\triangle$PQT is similar to $\triangle$PRS.
Find the length of QR (hint: find PR first).

2. These containers are mathematically similar. The volume of the large container is $2 \cdot 7$ litres. Calculate the volume of the smaller container.

20 cm   30 cm

3. A company sells a packet of porridge oats for £2·40 and plan to produce a similarly shaped packet half the height selling at £1·20. Is this smaller packet a good buy?

Trig – a review of basics

SOHCAHTOA

The sides of a right-angled triangle are named from the viewpoint of one of the angles.

The hypotenuse (the largest side) is always opposite the right-angle.

S O H   C A H   T O A

Use $\boxed{\sin}$ if Opp and Hyp are known or required.

Use $\boxed{\cos}$ if Adj and Hyp are known or required.

Use $\boxed{\tan}$ if Opp and Adj are known or required.

TOP TIP

Your calculator should show D or DEG not R or RAD or G or GRAD when working with angles. If not then change MODE to DEGREE.

Right-angled triangles – finding a side

Example

From the viewpoint of the 70° angle, the sides are named:

opp ✓   adj x hyp ✓

S O H   C A H   T A

Use $\boxed{\sin}$ and $\frac{\text{Opp}}{\text{Hyp}}$

So $\sin 70° = \frac{x}{10}$ (opposite) (multiply both (hypotenuse) sides by 10.)

$10 \sin 70° = x$

$\boxed{1}\boxed{0}\boxed{\times}\boxed{\sin}\boxed{7}\boxed{0}\boxed{=}$

$x = 9 \cdot 396\ldots \doteq 9 \cdot 40$ cm (to 3 s.f.)

Example

From the viewpoint of the 62° angle, the sides are named:

opp ✓   adj ✓   hyp x

S O H   C A H   T O A

Use $\boxed{\tan}$ and $\frac{\text{Opp}}{\text{Adj}}$

So $\tan 62° = \frac{8}{x}$ (opposite) (multiply both (adjacent) sides by x.)

$x \tan 62° = 8$ (divide both sides by $\tan 62°$.)

So $x = \frac{8}{\tan 62°}$

$\boxed{8}\boxed{\div}\boxed{\tan}\boxed{6}\boxed{2}\boxed{=}$

$x = 4 \cdot 253\ldots \doteq 4 \cdot 25$ cm (to 3 s.f.)

Right-angled triangles – finding an angle

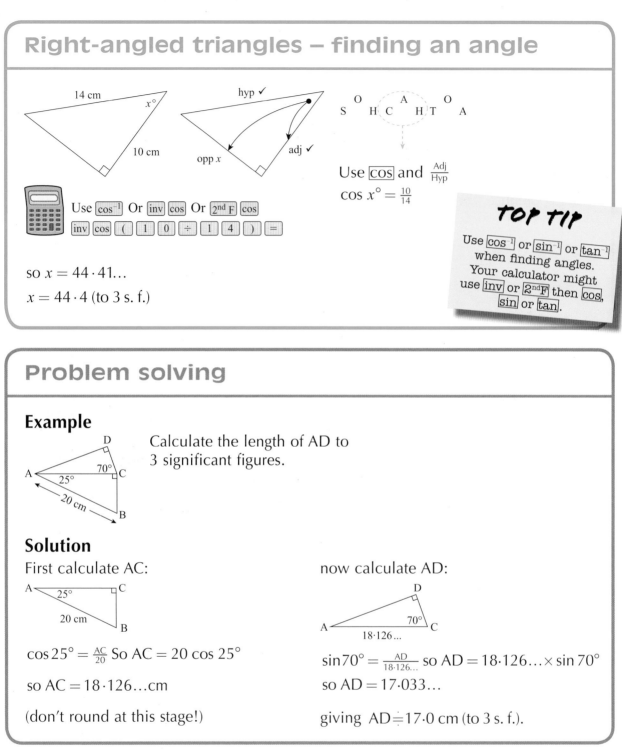

14 cm

$x°$

10 cm

hyp ✓

opp x

adj ✓

S $\underset{H\;\;C}{O}$ $\underset{C\;\;H}{A}$ $\underset{T\;\;A}{O}$

Use $\boxed{\cos}$ and $\frac{\text{Adj}}{\text{Hyp}}$

$\cos x° = \frac{10}{14}$

Use $\boxed{\cos^{-1}}$ Or $\boxed{\text{inv}}\,\boxed{\cos}$ Or $\boxed{2^{nd}\,F}\,\boxed{\cos}$

$\boxed{\text{inv}}\,\boxed{\cos}\,\boxed{(}\,\boxed{1}\,\boxed{0}\,\boxed{\div}\,\boxed{1}\,\boxed{4}\,\boxed{)}\,\boxed{=}$

so $x = 44 \cdot 41...$

$x = 44 \cdot 4$ (to 3 s. f.)

TOP TIP

Use $\boxed{\cos^{-1}}$ or $\boxed{\sin^{-1}}$ or $\boxed{\tan^{-1}}$ when finding angles. Your calculator might use $\boxed{\text{inv}}$ or $\boxed{2^{nd}F}$ then $\boxed{\cos}$, $\boxed{\sin}$ or $\boxed{\tan}$.

Problem solving

Example

D

$70°$ C

A

$25°$

20 cm

B

Calculate the length of AD to 3 significant figures.

Solution

First calculate AC:

A $\overset{25°}{}$ C

20 cm

B

$\cos 25° = \frac{AC}{20}$ So AC $= 20\cos 25°$

so AC $= 18 \cdot 126...$cm

(don't round at this stage!)

now calculate AD:

D

A $70°$ C

$18 \cdot 126 ...$

$\sin 70° = \frac{AD}{18 \cdot 126...}$ so AD $= 18 \cdot 126... \times \sin 70°$

so AD $= 17 \cdot 033...$

giving AD $\doteq 17 \cdot 0$ cm (to 3 s. f.).

Angles greater than 90°

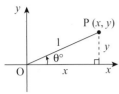

OP = 1 unit

Using SOHCAHTOA gives

$\sin \theta° = \frac{y}{1} = y$ and $\cos \theta° = \frac{x}{1} = x$

and $\tan \theta° = \frac{y}{x}$

Note: $\tan \theta° = \frac{\sin \theta°}{\cos \theta°}$

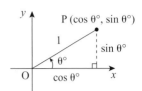

So if OP = 1 unit then:
x-coordinate of P gives $\cos \theta°$
y-coordinate of P gives $\sin \theta°$.

When $\theta°$ is greater than 90° the x- and y-coordinates of P still give the values of $\cos \theta°$ and $\sin \theta°$ and $\tan \theta°$ is still given by $\frac{\sin \theta°}{\cos \theta°}$.

This is a **unit** circle.
It has a radius of 1 unit.

Examples
Use your calculator to find:

(a) cos 123°

(b) tan 224° to 3 significant figures

Solutions

(a) $\cos 123° = -0\cdot5446... \doteq -0.545$
   (to 3 s.f.)

Note: Point P has a negative x-coordinate for 123°.

(b) $\tan 224° = 0\cdot9656... \doteq 0\cdot966$
   (to 3 s.f.)

TOP TIP

If $\theta° = 90°$ then P(0, 1) so cos 90° = 0, sin 90° = 1.

Quick Test 29

1. For each triangle calculate x correct to 3 significant figures.

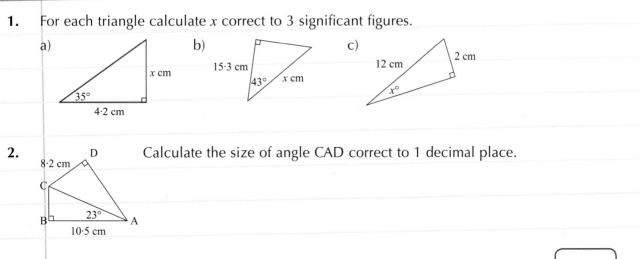

a) [triangle with 35°, 4·2 cm, x cm]

b) [triangle with 15·3 cm, 43°, x cm]

c) [triangle with 12 cm, 2 cm, x°]

2. Calculate the size of angle CAD correct to 1 decimal place.

[triangle D, 8·2 cm, C, B, 23°, A, 10·5 cm]

Working with trig graphs

The unit circle diagram

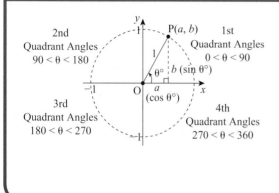

2nd
Quadrant Angles
$90 < \theta < 180$

1st
Quadrant Angles
$0 < \theta < 90$

3rd
Quadrant Angles
$180 < \theta < 270$

4th
Quadrant Angles
$270 < \theta < 360$

From a starting position along the x-axis, line OP rotates about the origin O anticlockwise $\theta°$, as shown in the diagram. Here are three definitions:

$$\sin \theta° = b \qquad \cos \theta° = a \qquad \tan \theta° = \frac{b}{a}$$

Since a and b are coordinates they may be positive or negative.

Since OP $= 1$ unit P sweeps out a **UNIT CIRCLE**.

The sine, cosine and tangent graphs

You can draw graphs showing the values of these three **trigonometric functions** $\sin \theta°$, $\cos \theta°$ and $\tan \theta°$ as OP rotates from $0°$ through all 4 quadrants to $360°$.

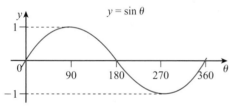

$$y = \sin \theta$$

$\sin 0° = 0 \quad \sin 90° = 1 \quad \sin 180° = 0 \quad \sin 270° = -1 \quad \sin 360° = 0$

$\sin \theta°$ is positive in the 1st and 2nd quadrants.

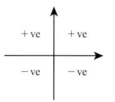

$+$ ve $\quad +$ ve

$-$ ve $\quad -$ ve

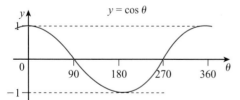

$$y = \cos \theta$$

$\cos 0° = 1 \quad \cos 90° = 0 \quad \cos 180° = -1 \quad \cos 270° = 0 \quad \cos 360° = 1$

$\cos \theta°$ is positive in the 1st and 4th quadrants.

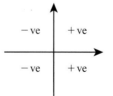

$-$ ve $\quad +$ ve

$-$ ve $\quad +$ ve

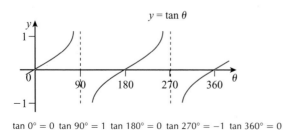

$$y = \tan \theta$$

$\tan 0° = 0 \quad \tan 90° = 1 \quad \tan 180° = 0 \quad \tan 270° = -1 \quad \tan 360° = 0$
is undefined $\qquad$ is undefined

$\tan \theta°$ is positive in the 1st and 3rd quadrants.

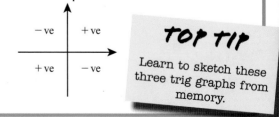

$-$ ve $\quad +$ ve

$+$ ve $\quad -$ ve

TOP TIP

Learn to sketch these three trig graphs from memory.

The quadrant diagram

TOP TIP

Use this quadrant diagram to solve equations, see P. 96.

Summary diagram:

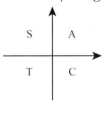

1st quadrant: $\sin\theta°$, $\cos\theta°$, $\tan\theta°$ are All positive **(A)** $(0<\theta<90)$.

2nd quadrant: only **Sin** $\theta°$ is positive **(S)** $(90<\theta<180)$.

3rd quadrant: only **Tan** $\theta°$ is positive **(T)** $(180<\theta<270)$.

4th quadrant: only **Cos** $\theta°$ is positive **(C)** $(270<\theta<360)$.

Solving trig equations using the trig graphs

Special trig equations like:

$\sin x° = 1$, $\sin x° = 0$, $\sin x° = -1$, $\cos x° = 1$, $\cos x° = 0$, $\cos x° = -1$ and $\tan x° = 0$

can be solved by reading the solutions from the trig graphs.

For example if $0 \le x \le 360$ $\sin x° = 0$ has three solutions:

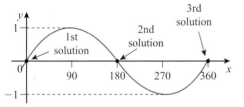

The solutions are $x = 0, 180, 360$

Example
Solve $\cos x° = 1$ for $0 \le x \le 360$

Solution

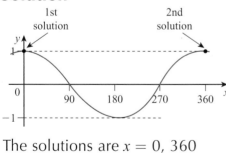

The solutions are $x = 0, 360$

Example
Solve $\sin x° = 2$ for $0 \le x \le 360$

Solution
From the sine graph $\sin x°$ has a maximum value of 1 so there is no value of x for which $\sin x° = 2$. This equation has no solution.

Quick Test 30

1. Sketch the graphs $y = \sin x°$, $y = \cos x°$ and $y = \tan x°$ for $0 \le x \le 360$ (make sure the scales on both axes are clearly indicated).

2. Solve the following equations for $0 \le x \le 360$:

   a) $\sin x° = -1$ b) $\tan x° = 0$ c) $\cos x° = -1$

Working with related trig graphs

The graphs $y = k \sin bx° + c$ and $y = k \cos bx° + c$

The **amplitude** is half the difference between the maximum and minimum values on the graph.

The **period** is the 'length' of 1 cycle measured in units along the x-axis.

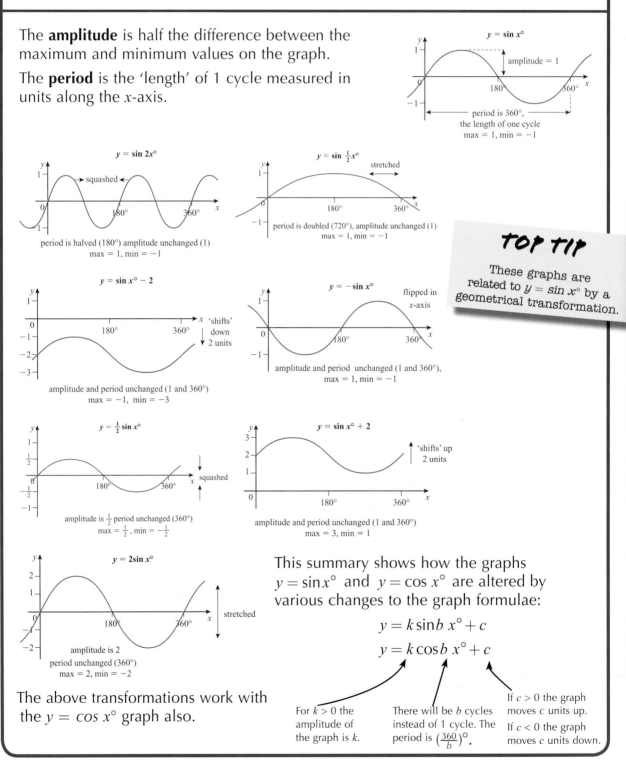

$y = \sin x°$

amplitude = 1

period is 360°, the length of one cycle
max = 1, min = −1

$y = \sin 2x°$

squashed

period is halved (180°) amplitude unchanged (1)
max = 1, min = −1

$y = \sin \frac{1}{2}x°$

stretched

period is doubled (720°), amplitude unchanged (1)
max = 1, min = −1

$y = \sin x° − 2$

'shifts' down 2 units

amplitude and period unchanged (1 and 360°)
max = −1, min = −3

$y = −\sin x°$

flipped in x-axis

amplitude and period unchanged (1 and 360°),
max = 1, min = −1

$y = \frac{1}{2} \sin x°$

squashed

amplitude is $\frac{1}{2}$ period unchanged (360°)
max = $\frac{1}{2}$, min = $−\frac{1}{2}$

$y = \sin x° + 2$

'shifts' up 2 units

amplitude and period unchanged (1 and 360°)
max = 3, min = 1

$y = 2\sin x°$

stretched

amplitude is 2
period unchanged (360°)
max = 2, min = −2

TOP TIP

These graphs are related to $y = \sin x°$ by a geometrical transformation.

This summary shows how the graphs $y = \sin x°$ and $y = \cos x°$ are altered by various changes to the graph formulae:

$$y = k \sin b\ x° + c$$
$$y = k \cos b\ x° + c$$

The above transformations work with the $y = \cos x°$ graph also.

For $k > 0$ the amplitude of the graph is k.

There will be b cycles instead of 1 cycle. The period is $\left(\frac{360}{b}\right)°$.

If $c > 0$ the graph moves c units up.

If $c < 0$ the graph moves c units down.

The graphs $y = \sin(x + d)°$ and $y = \cos(x + d)°$

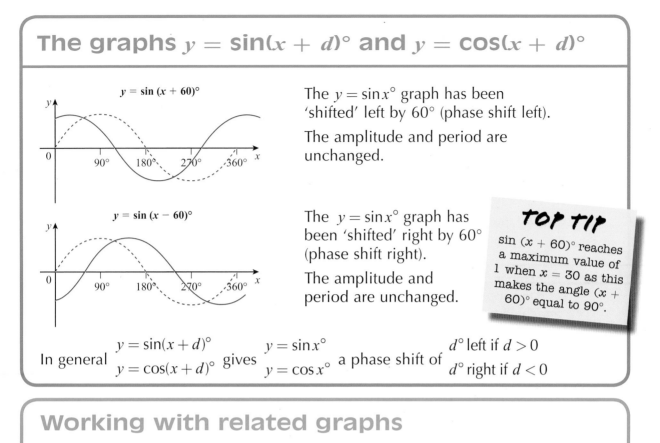

$y = \sin (x + 60)°$

The $y = \sin x°$ graph has been 'shifted' left by 60° (phase shift left).

The amplitude and period are unchanged.

$y = \sin (x - 60)°$

The $y = \sin x°$ graph has been 'shifted' right by 60° (phase shift right).

The amplitude and period are unchanged.

TOP TIP

$\sin (x + 60)°$ reaches a maximum value of 1 when $x = 30$ as this makes the angle $(x + 60)°$ equal to 90°.

In general $\begin{array}{l} y = \sin(x + d)° \\ y = \cos(x + d)° \end{array}$ gives $\begin{array}{l} y = \sin x° \\ y = \cos x° \end{array}$ a phase shift of $\begin{array}{l} d° \text{ left if } d > 0 \\ d° \text{ right if } d < 0 \end{array}$

Working with related graphs

The diagram shows the graph $y = k \sin ax°$

Examples

(a) State the values of k and a.

(b) Give the maximum and minimum values of $k \sin ax°$.

(c) State the period of the graph.

Solutions

(a) The amplitude is 5 so $k = 5$. From the x-axis scale there will be 4 cycles from 0° to 360° so $a = 4$.

(b) The maximum is 5 and the minimum is −5. (c) The period is 90°.

Quick Test 31

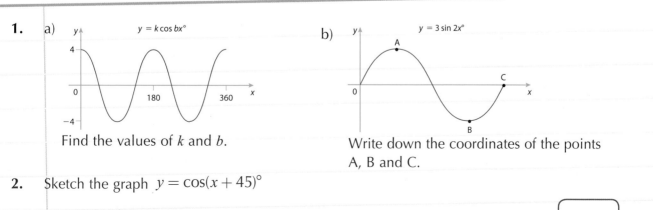

1. a) $y = k \cos bx°$

Find the values of k and b.

b) $y = 3 \sin 2x°$

Write down the coordinates of the points A, B and C.

2. Sketch the graph $y = \cos(x + 45)°$

Working with trig equations

Solving trig equations using the quadrant diagram

Step 1   Rearrange the equation (if possible!) to the form:

$$\sin(\text{angle}) = \text{number}$$
$$\text{or } \cos(\text{angle}) = \text{number}$$
$$\text{or } \tan(\text{angle}) = \text{number}$$

Example

Solve $5\sin x° + 1 = 0$, $0 \le x \le 360$

Solution

Step 1 $5\sin x° = -1$

$$\sin x° = -\frac{1}{5}$$

$$\sin x° = -0 \cdot 2$$

Step 2   Is 'number' positive or negative? Use the answer to this question and the quadrant diagram below to find which quadrants 'angle' is in.

Step 2 $-0 \cdot 2$ is negative

Using the diagram:

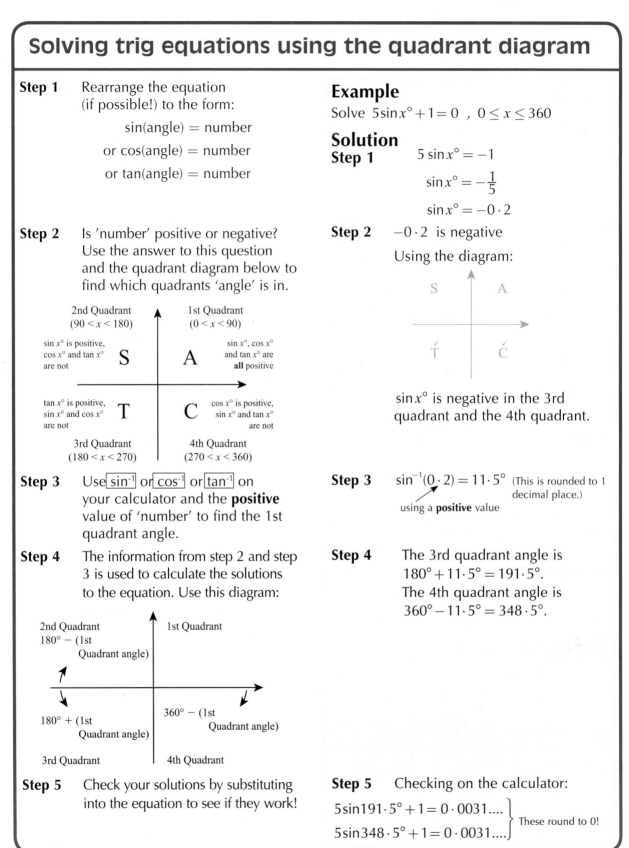

$\sin x°$ is negative in the 3rd quadrant and the 4th quadrant.

Step 3   Use $\boxed{\sin^{-1}}$ or $\boxed{\cos^{-1}}$ or $\boxed{\tan^{-1}}$ on your calculator and the **positive** value of 'number' to find the 1st quadrant angle.

Step 3 $\sin^{-1}(0 \cdot 2) = 11 \cdot 5°$ (This is rounded to 1 decimal place.)

using a **positive** value

Step 4   The information from step 2 and step 3 is used to calculate the solutions to the equation. Use this diagram:

Step 4   The 3rd quadrant angle is $180° + 11 \cdot 5° = 191 \cdot 5°$.
The 4th quadrant angle is $360° - 11 \cdot 5° = 348 \cdot 5°$.

Step 5   Check your solutions by substituting into the equation to see if they work!

Step 5   Checking on the calculator:

$5\sin 191 \cdot 5° + 1 = 0 \cdot 0031....$
$5\sin 348 \cdot 5° + 1 = 0 \cdot 0031....$ } These round to 0!

Solutions to equations using trig graphs

The solutions to the equations

$\sin x° = \frac{1}{2}$ and $\sin x° = -\frac{1}{2}$

for $0 \le x \le 360$ are shown

in the diagram on the right.

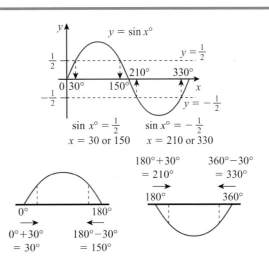

Using 0°, 180° and 360° these
solutions can be calculated
using the symmetry of the graph
and knowing $\sin^{-1}\left(\frac{1}{2}\right) = 30°$.

This symmetry can be used to solve trig equations as an alternative to using the
quadrant diagram method or as a way of checking that your solutions make sense.

Example

Solve $4\cos x° + 3 = 0$ for $0 \le x \le 360$.

Solution

$4\cos x° = -3$ so $\cos x° = -\frac{3}{4}$

now $\cos^{-1}\left(\frac{3}{4}\right) = 41·4°$

From the graph:

$x = 180 - 41·4 = 138·6$ (using point A)

or $x = 180 + 41·4 = 221·4$ (using point B)

Note: The coordinates of A are $(138·6, -0·75)$ and
B are $(225·4, -0·75)$

TOP TIP

Always use the trig graphs
to solve equations like
$\sin x° = 0$, $\cos x° = -1$,
$\tan x° = 0$, $\sin x° = 1$, etc.
Solutions involve
0°, 90°, 180°, 270°, 360°.

Quick Test 32

1. Solve these equations for $0 \le x \le 360$:

   a) $3\sin x° - 2 = 0$ b) $2 + 5\cos x° = 1$

2. The diagram shows the graph
   $y = \sin x°$. The line $y = -0·4$
   cuts the graph at P and Q.
   Find the coordinates of P and Q.

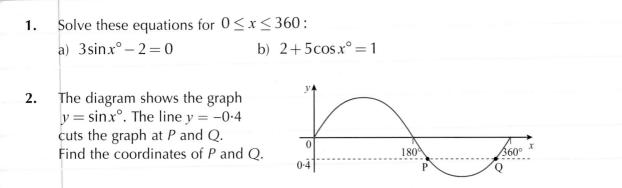

Some trig formulae and applications

TOP TIP

The graphs here are related to the sine and cosine graphs on P 94.

Applications of trig formulae

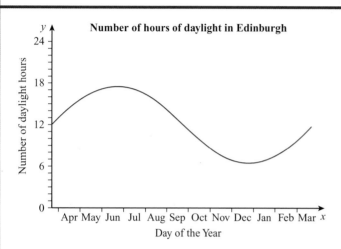

Number of hours of daylight in Edinburgh

Number of daylight hours

Apr May Jun Jul Aug Sep Oct Nov Dec Jan Feb Mar x
Day of the Year

This graph shows the number of daylight hours in Edinburgh over the course of a year (mid-March to mid-March).

It shows a typical 'sinusoidal' shape, i.e the shape of a sine curve or a cosine curve.

This situation and others, e.g. water depth at a harbour, height of a bouncing object on a spring, have a similar naturally occurring cyclic behaviour.

These situations can often be described mathematically (modelled) using a trig formula like:

$y = a \sin bx° + c$ or $y = a \cos bx° + c$

The example below illustrates how this is done:

Example

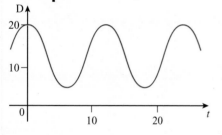

The depth of water, D metres, at a harbour, t hours after midnight, is given by the formula:

$D = 8·5\cos(30t)° + 13$

(a) Find the depth of the water at 3 p.m.

(b) Find the difference between the maximum and minimum depths over time.

Solution

(a) 3 p.m. is 15 hours after midnight so let $t = 15$. This gives
$D = 8·5 \cos(30×15)° + 13 = 8·5 \cos 450° + 13 = 13$ (since cos 450° = 0).
At 3 p.m. the depth is therefore 13 metres.

(b) The value of $\cos(30t)°$ ranges from 1 to −1 as the angle $30t°$ changes value.

The maximum depth is $8·5×1 + 13 = 21·5$ m, the minimum is $8·5×(−1) + 13 = 4·5$ m.

So the difference is $21·5 − 4·5 = 17$ metres.

Some trig formulae

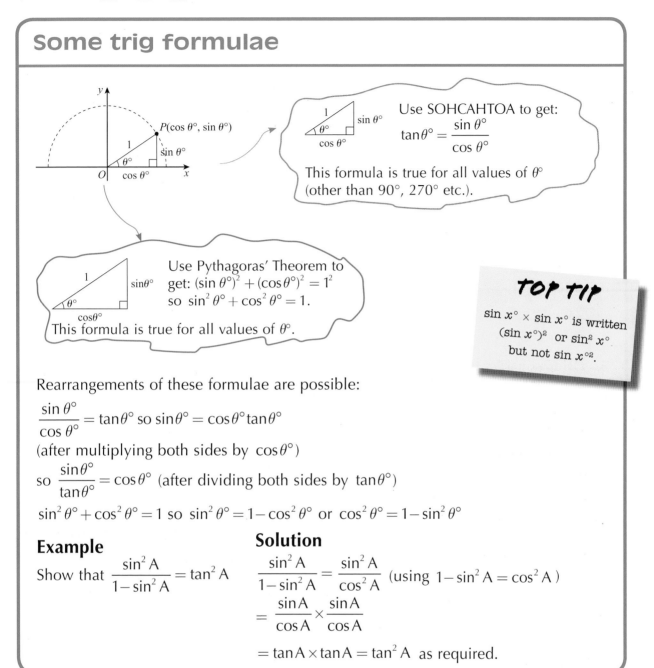

Use SOHCAHTOA to get:
$$\tan\theta° = \frac{\sin\theta°}{\cos\theta°}$$

This formula is true for all values of $\theta°$ (other than 90°, 270° etc.).

Use Pythagoras' Theorem to get: $(\sin\theta°)^2 + (\cos\theta°)^2 = 1^2$
so $\sin^2\theta° + \cos^2\theta° = 1$.

This formula is true for all values of $\theta°$.

TOP TIP

$\sin x° \times \sin x°$ is written $(\sin x°)^2$ or $\sin^2 x°$ but not $\sin x°^2$.

Rearrangements of these formulae are possible:

$\dfrac{\sin\theta°}{\cos\theta°} = \tan\theta°$ so $\sin\theta° = \cos\theta°\tan\theta°$

(after multiplying both sides by $\cos\theta°$)

so $\dfrac{\sin\theta°}{\tan\theta°} = \cos\theta°$ (after dividing both sides by $\tan\theta°$)

$\sin^2\theta° + \cos^2\theta° = 1$ so $\sin^2\theta° = 1 - \cos^2\theta°$ or $\cos^2\theta° = 1 - \sin^2\theta°$

Example

Show that $\dfrac{\sin^2 A}{1 - \sin^2 A} = \tan^2 A$

Solution

$$\frac{\sin^2 A}{1 - \sin^2 A} = \frac{\sin^2 A}{\cos^2 A} \quad \text{(using } 1 - \sin^2 A = \cos^2 A\text{)}$$
$$= \frac{\sin A}{\cos A} \times \frac{\sin A}{\cos A}$$
$$= \tan A \times \tan A = \tan^2 A \text{ as required.}$$

Quick Test 33

1. The depth of water, D metres, at a harbour, t hours after noon is given by the formula
    $D = 14{\cdot}6 + 7{\cdot}9\sin(30t)°$.

    a) Find the depth of the water at 1 a.m.

    b) Find the maximum and minimum depths of the water during the course of a day and the times they occur.

2. Show that:

    a) $1 - 2\sin^2 A° = \cos^2 A° - \sin^2 A°$ b) $\dfrac{1 - \cos^2\theta°}{\cos^2\theta°} = \tan^2\theta°$

Sample unit 2 test questions

Linear equations, etc.

1. Find the equation of the straight line passing through the point $(2, -1)$ with gradient -1.

2. Solve the inequation $5m + 3 > m - 5$.

3. Poflake Porridge Oats are sold in two different sized packets: large and small.
   In total 3 large packets and 4 small packets weigh $2 \cdot 58$ kg.
   Write down an algebraic equation to illustrate this.

4. Solve algebraically this system of equations:
   $2x - 3y = 7$
   $x + y = 6$

5. Using suitable units of measurement, the pressure P of water on a diver at depth D is given by the formula:
   $P = 15 + \frac{1}{2}D$

   Change the subject of the formula to D.

Quadratic graphs, etc.

1.

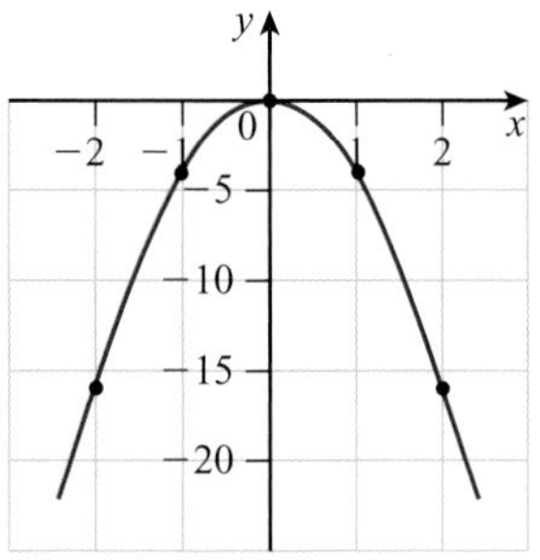

DIAGRAM 1

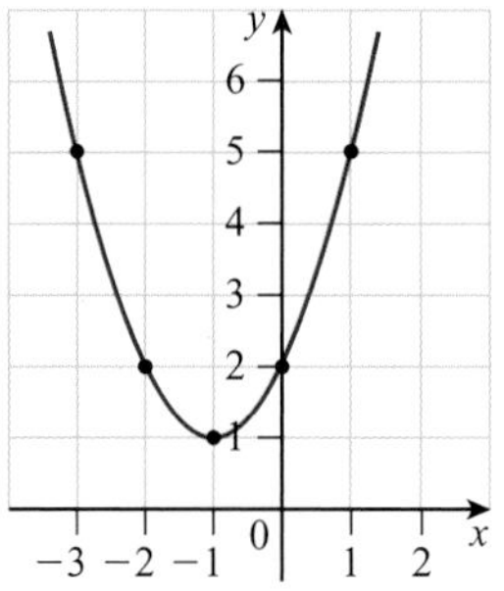

DIAGRAM 2

(a) The graph shown in diagram 1 has formula $y = kx^2$. Find the value of k.

(b) The graph shown in diagram 2 has formula $y = (x + b)^2 + c$. Find the values of b and c.

2. For the graph with equation $y = (x + 2)^2 - 3$ write down:

(a) The equation of the axis of symmetry.

(b) The coordinates of the minimum turning point.

3. Sketch the graph $y = (x - 2)(x - 4)$ showing clearly where it crosses the x-axis and the y-axis. Also show clearly the coordinates of the turning point.

Quadratic equations, etc.

1. Use the quadratic formula to solve $x^2 + 7x + 2 = 0$

2. Solve the equation $(x + 9)(x - 1) = 0$

3. Calculate the discriminant of the equation $5x^2 - x + 1 = 0$ and say what this tells you about the nature of the roots of this equation.

Angles and similarity, etc.

1. Do three rods measuring 14·8 cm, 25·9 cm and 30·1 cm make a right-angled triangle when placed end-to-end? (Show clearly your reasoning.)

2. Two tangent lines from Q to the circle centre C touch the circle at P and R and form an angle of 43°. Determine the size of angle PCR.

3. These two cups are mathematically similar. The larger cup has height 20 cm and holds 225 ml. How much does the smaller cup hold if it has a height of 16 cm?

4. This wooden puzzle has three large pieces that are regular heptagons (seven-sided shapes) and three smaller quadrilaterals.

   Calculate the shaded angle indicated on one of the larger pieces in the diagram.

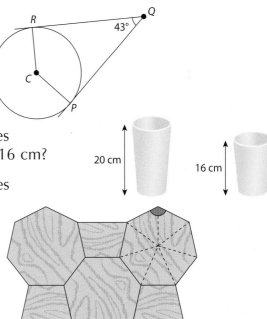

Trig graphs and equations

1. Solve the equation $3 \sin x° - 2 = 0$ for $0 \le x \le 360$.

2. Sketch the graph $y = \sin x° + 1$ for $0 \le x \le 360$.

3. Write down the amplitude and period of the graph: $y = 3 \cos 2 x°$.

Sample end-of-course exam questions on unit 2 topics

Non-calculator questions

1. $g(x) = x^2 - 2x$. Evaluate $g(-3)$.

2. The diagram shows part of the graph $y = 4 \sin 2x°$. Write down the coordinates of the points A and B.

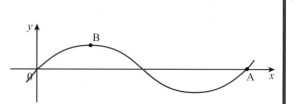

3.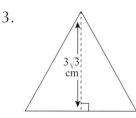
An equilateral triangle has a height of exactly $3\sqrt{3}$ cm, as shown in the diagram.
Calculate the exact length of its sides.

4. Part of the graph $y = kx(x - 5)$, where k is a constant, is shown in the diagram. The point A (2, 18) lies on the graph. Calculate the value of k.

5. Solve the inequality $3(5 - x) \leq 5x - 1$.

6.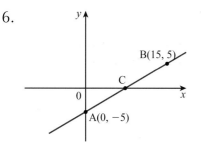
The diagram shows a line passing through A (0, −5) and B (15, 5).

   (a) Find the equation of the line.

   (b) Find the coordinates of the point C where the line cuts the x-axis.

7. Maureen and Ken are buying bread and rolls.

   (a) Maureen buys 5 loaves and half a dozen rolls at a cost of £5·81. Write down an algebraic equation to illustrate this.

   (b) Ken's bill comes to £3·59. He bought 3 loaves and 4 rolls. Write down an algebraic equation to illustrate this.

   (c) Find the total cost of a dozen rolls and 4 loaves.

Calculator-allowed questions

1. The diagram shows the largest cross-section of a spheri-
   cal glass paperweight with radius 3·4 cm. The base of the
   paperweight has diameter 3·1 cm.

   Calculate the height, h cm, of the paperweight.

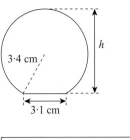

3·4 cm

3·1 cm

2. Solve algebraically the equation:
   $2 \tan x° + 2 = \sin 80°$ $0 \le x < 360$

3. A symmetrical lawn consists of two identical rectangles, as
   shown. The breadth of each rectangle is x metres and has
   length twice its breadth. A path, 1 m wide, surrounds the
   lawn on all sides.

   (a) If the total area of the lawn is 20 m² more than the
       area of the path, show that:
       $4x^2 - 10x - 24 = 0.$

   (b) Hence find the dimensions of each of the rectangular
       pieces of the lawn.

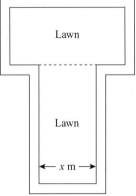

Lawn

Lawn

x m

4.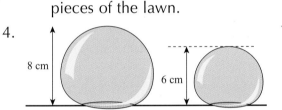

8 cm

6 cm

   These two glass
   paperweights are mathematically similar
   in shape. Their heights are 8 cm and
   6 cm. If the volume of the larger
   paperweight is 372 cm³, calculate the
   volume of the smaller weight to
   3 significant figures.

5. A ramp is constructed from two sections. The cross-section of the ramp
   is shown in the diagram with section 1 forming a right-angled triangle
   ABF. Section 2 consists of a rectangle BDEF surmounted by right-angled
   triangle BCD.

   AB = 6 m, AF = 5·8 m, FE = 6·5 m and CE = 4 m.

   C

   B

   6 m

   Section 2

   D 4 m

   Section 1

   A

   5·8 m

   F

   6·5 m

   E

   Calculate angle DBC, the angle of incline of section 2 from the horizontal,
   to 1 decimal place.

Area of a triangle

Naming sides and angles of triangles

There is a system (convention) used for naming the angles and sides in a triangle.

The small letters used for the sides match the capital letters used for the opposite angles, as shown in this diagram.

Here are other examples:

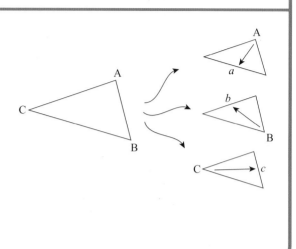

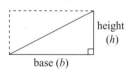

Area of a triangle (non-trig!)

If the triangle is right-angled:

Area $= \frac{1}{2}bh$.

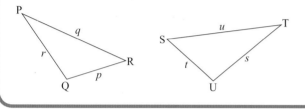

Or, if you know the height:

Area $= \frac{1}{2}bh$.

> **TOP TIP**
>
> If your triangle has a right-angle then don't use the trig formula to find its area!

Trig formula for the area of a triangle

If you know two sides and the angle included between these sides then you can find the area of the triangle:

Area $= \frac{1}{2}\,ab\sin C$

Area $= \frac{1}{2}\,ac\sin B$

Area $= \frac{1}{2}\,bc\sin A$

> **TOP TIP**
>
> Always check your calculator is in degree mode (DEG or D) before using $\boxed{\sin^{-1}}$.

Example

80 cm
32°
90 cm

Calculate the area of this triangular flag.

Solution

Area $= \frac{1}{2} \times 80 \times 90 \times \sin 32°$

$= 1907 \cdot 70\ldots$

$\doteqdot \mathbf{1910\ cm^2}$ (to 3 s.f.)

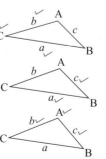

Sin θ° = *k* where *k* is positive

Example
Solve sin $\theta° = 0.2$ $0 < \theta < 180$

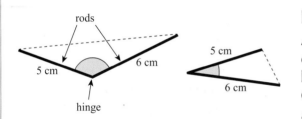

TOP TIP

When solving triangle problems remember that solving the sine of an angle equals a positive number may have two possible angles (one in the 1st quadrant and one in the 2nd quadrant).

Solution

sin $\theta°$ is positive in the 1st and 2nd quadrants. Use your calculator to get the 1st quadrant angle: sin^{-1} $0.2 \doteqdot 11.5°$. The 2nd quadrant angle is $180° - 11.5° = 168.5°$.

So sin $\theta° = 0.2$ has two possible solutions $11.5°$ and $168.5°$.

Two possible angles

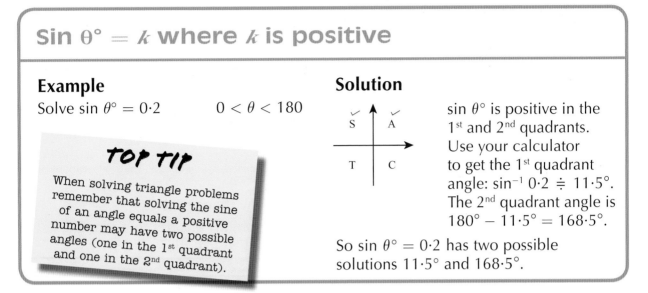

Example
As shown in the diagrams, there are two different positions that the two rods can be in to form a triangle with area 5 cm². Calculate the two possible sizes of the angle between the rods to 1 decimal place.

Solution
Use Area $= \frac{1}{2}ab$ sin C with Area $= 5$, $a = 5$, $b = 6$

so $5 = \frac{1}{2} \times 5 \times 6 \times$ sin C $\Rightarrow 5 = 15 \times$ sin C $\Rightarrow$ sin C $= \frac{5}{15} = \frac{1}{3}$

> sin$^{-1}\left(\frac{1}{3}\right) = 19.47°...$ gives the 1st quadrant angle.

so $\angle C = 19.47°...$ or $180° - 19.47°... = 160.52°...$

The two possible sizes for the angle are $19.5°$ and $160.5°$.

Quick Test 34

1. Find the area of each triangle. Give your answer to 3 significant figures.

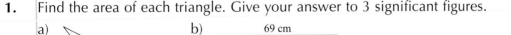

a)   b)

23·6 cm   69 cm

11·5 cm   37°

                                     45 cm

                                     61°

2. Triangle ABC has area of 16 cm². AB is 4 cm and AC is 12 cm. Calculate possible sizes of angle BAC to 1 decimal place.

The Sine Rule

What is the Sine Rule?

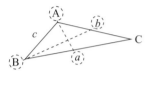

The Sine Rule is the following result:

$$\frac{a}{\sin A} = \frac{b}{\sin B} = \frac{c}{\sin C}$$

This is true for any triangle ABC.

The result helps you find missing sizes of angles or lengths of sides in a triangle. Problems where the Sine Rule is useful can be identified like this:

Your problem will involve two pairs of 'opposites':

If any three of $\angle A$, $\angle B$, a and b are known then the fourth can be calculated using the Sine Rule.

Similarly for knowing any three of $\angle A$, $\angle C$, a and c or knowing any three of $\angle B$, $\angle C$, b and c you can calculate the fourth.

TOP TIP

Remember if you know two angles in a triangle you can easily calculate the third!

Finding a side

Example

In triangle PQR, $\angle PQR = 135°$, $\angle QRP = 32°$ and PR = 8·2 cm. Calculate the length of side PQ to 3 significant figures.

Solution

Step 1 Label the sides and angles that you know or are trying to calculate.

Step 2 Write down the Sine Rule and identify the part you will use.

$$\frac{p}{\sin P} = \frac{q}{\sin Q} = \frac{r}{\sin R} \quad \text{use} \quad \frac{q}{\sin Q} = \frac{r}{\sin R}$$

Step 3 Substitute and solve:

$$\frac{8·2}{\sin 135°} = \frac{r}{\sin 32°} \Rightarrow \frac{8·2 \times \sin 32°}{\sin 135°} = \frac{r \times \sin 32°}{\sin 32°}$$

(multiplying both sides by sin 32°)

So $r = \dfrac{8·2 \times \sin 32°}{\sin 135°} = 6·145...$ PQ $\doteqdot 6·15$ cm (to 3 significant figures).

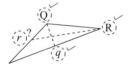

Finding an angle

Example

The 20-metre top support on a crane jib makes a 120° angle with the main crane tower, as shown in the diagram.

Calculate $x°$, the angle the lower support, which is 26 metres long, makes with the tower. Give your answer to 1 decimal place.

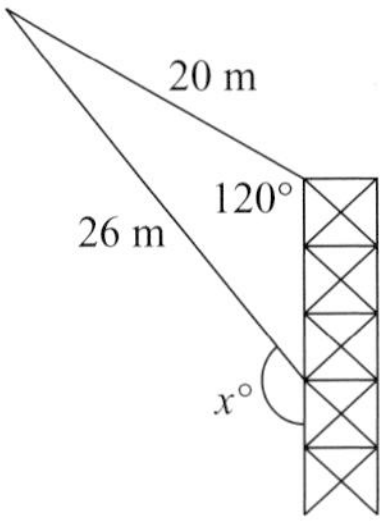

Solution

Use $\dfrac{b}{\sin B} = \dfrac{c}{\sin C}$ giving $\dfrac{26}{\sin 120°} = \dfrac{20}{\sin C} \Rightarrow \dfrac{26 \times \sin C}{\sin 120°} = \dfrac{20 \times \sin C}{\sin C}$

so $\dfrac{26 \sin C}{\sin 120°} = 20 \Rightarrow \dfrac{26 \sin C \times \sin 120°}{\sin 120°} = 20 \times \sin 120° \Rightarrow 26 \sin C = 20 \times \sin 120°$

$\Rightarrow \dfrac{26 \sin C}{26} = \dfrac{20 \sin 120°}{26} \Rightarrow \sin C = \dfrac{20 \sin 120°}{26} = 0.6661\ldots$

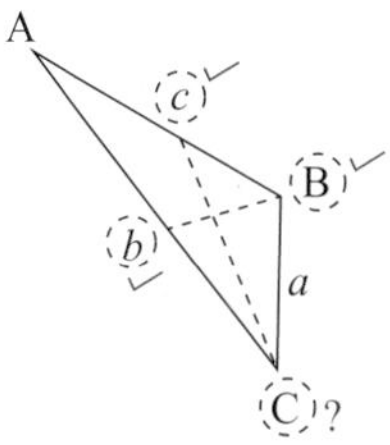

Now use $\sin^{-1} 0.6661\ldots$ to get $\angle C = 41.77°\ldots$

So $x° = 180° - 41.77°\ldots = 138.22°\ldots$ required angle is $138.2°$
(to 1 decimal place).

Two possible angles

Suppose triangle ABC has AB = 5 cm, $\angle A = 15°$ and side BC = 2 cm.

There are two possible positions for side BC.

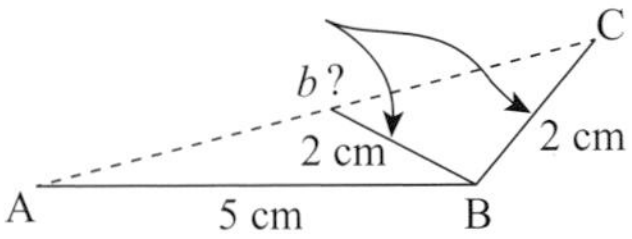

Here is the Sine Rule calculation for $\angle C$:

$\dfrac{2}{\sin 15°} = \dfrac{5}{\sin C} \Rightarrow \sin C = \dfrac{5 \sin 15°}{2} = 0.647\ldots$

So $\angle C \doteq 40.3°$ or $\angle C \doteq 180 - 40.3° = 139.7°$
(1st quadrant solution   (2nd quadrant solution).
from $\sin^{-1} 0.647\ldots$).

Quick Test 35

1. a) Calculate DE (to 3 s.f.).   b) Calculate angle P (to 1 decimal place).

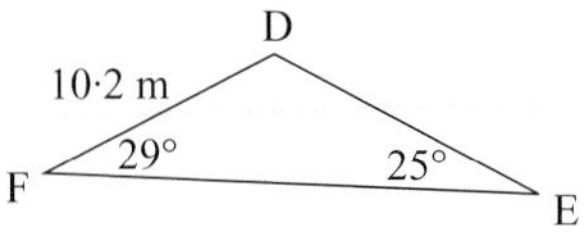

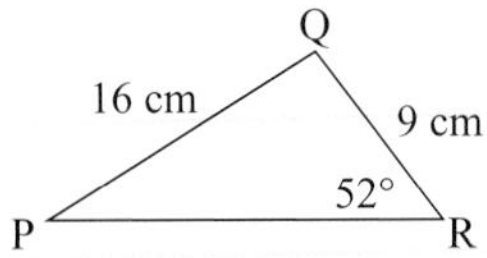

2. 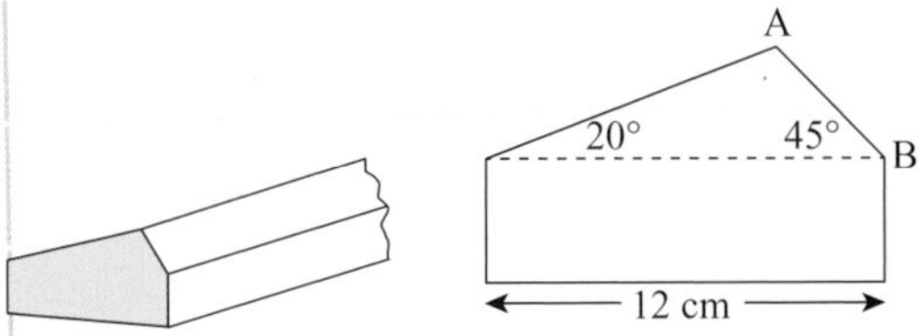 A wooden rod has a uniform cross-section in the shape of a rectangle surmounted by a triangle, as shown. Calculate the length AB to 3 significant figures.

The Cosine Rule

TOP TIP

If you know two sides and the angle in between then you can use the **Cosine Rule** to find the other side.

What is the Cosine Rule?

The Cosine Rule is the following result:

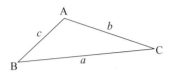

$$a^2 = b^2 + c^2 - 2bc \cos A$$
$$b^2 = a^2 + c^2 - 2ac \cos B$$
$$c^2 = a^2 + b^2 - 2ab \cos C$$

Three versions of the Cosine Rule for triangle ABC.

This rule is true for any triangle ABC.

These versions of the Cosine Rule allow you to calculate a missing side length.

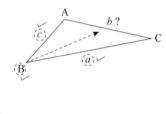

If you know two sides and the angle between them (the included angle) then you can use these results to find the side opposite the known angle, e.g. to find b in the diagram on the left use:

$$b^2 = a^2 + c^2 - 2ac \cos B$$

Finding a side

Example

In triangle RST ∠STR = 43°, RT = 12 cm and ST = 8 cm. Calculate the length of side RS to 3 significant figures.

Solution

Step 1 Label the sides and angles that you know or are trying to calculate.

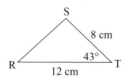

Step 2 Write down the version of the Cosine Rule you will use

$$t^2 = r^2 + s^2 - 2rs \cos T$$

Step 3 Substitute and solve:

$$t^2 = 8^2 + 12^2 - 2 \times 8 \times 12 \times \cos 43°.$$

So $t^2 = 67 \cdot 58...$ (now take the square root)

giving $t = \sqrt{67 \cdot 58...} = 8 \cdot 220...$

so side RS = $8 \cdot 22$ cm correct to 3 significant figures.

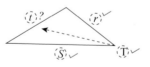

Rearranging the Cosine Rule

You can 'change the subject' of the Cosine Rule so that it is better suited to find a missing angle.

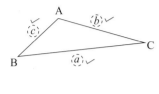

$$\cos A = \frac{b^2 + c^2 - a^2}{2bc}$$

$$\cos B = \frac{a^2 + c^2 - b^2}{2ac}$$

$$\cos C = \frac{a^2 + b^2 - c^2}{2ab}$$

Three versions of the Cosine Rule.

If you know the three sides a, b and c use these versions to find $\angle A$, $\angle B$ or $\angle C$.

Finding an angle

TOP TIP

For cos (angle) = negative number use $\boxed{\cos^{-1}}$ with the number ignoring the −ve sign. Then use 180° minus this '1st quadrant' angle.

Example

The triangle PQR has sides PQ = 8 cm, QR = 9 cm and PR = 7 cm.

Calculate the size of angle PQR to 1 decimal place.

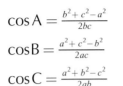

Calculator tip

Be careful with the division. It's safer to calculate:
$9^2 + 8^2 − 7^2 = 96$,
then $2 \times 9 \times 8 = 144$,
followed by the division
$96 \div 144 = 0.666...$

Solution

Use $\cos Q = \dfrac{p^2 + r^2 - q^2}{2pr}$

so $\cos Q = \dfrac{9^2 + 8^2 - 7^2}{2 \times 9 \times 8} = \dfrac{96}{144}$ giving $\cos Q = 0.666...$
so $\angle Q = 48.18... \doteq 48.2°$
(to 1 decimal place).

Quick Test 36

1. a) Calculate DF to 1 decimal place.   b) Calculate the size of angle BAC to 1 decimal place.

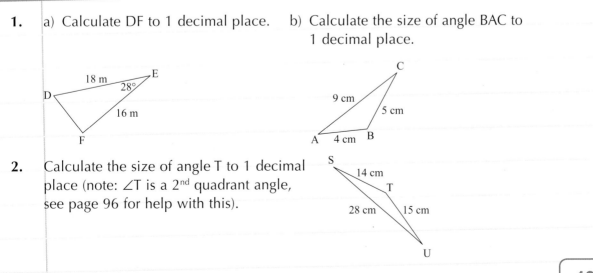

2. Calculate the size of angle T to 1 decimal place (note: $\angle T$ is a 2nd quadrant angle, see page 96 for help with this).

Bearings

TOP TIP

The point from which a bearing is taken needs a North line drawn.

What is a three-figure bearing?

From location A, the direction you need to travel to get to location B may be described by an angle. Here are the steps:

Step 1   Stand at A and face north.

Step 2   Turn clockwise to face B.

Step 3   Measure the angle through which you turned.

This angle is the **bearing** of B from A.

B lies on a bearing of 100° from A.

Note:

Three digits are always used to describe a bearing. So, for example, 028° is used for the angle 28°.

Example

The bearing of a lighthouse from a ship is 053°. Calculate the bearing of the ship from the lighthouse.

Solution

You must now stand at the lighthouse to take the required bearing:

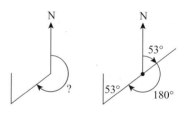

From the diagram, it can be seen that the required bearing is $53° + 180° = 233°$.

A problem involving the Cosine Rule

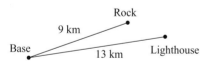

The diagram shows a helicopter base. The rock has a bearing of 070° from the base and the lighthouse has a bearing of 080° from the base. How far is the lighthouse from the rock?

Solution

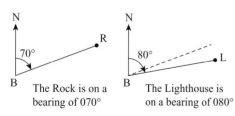

The Rock is on a bearing of 070°

The Lighthouse is on a bearing of 080°

∠LBR = 80° − 70° = 10°
This is the difference of the bearings

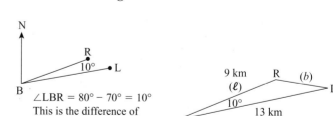

Now use the Cosine Rule in triangle BRL:

$b^2 = r^2 + \ell^2 - 2r\ell \cos B = 13^2 + 9^2 - 2 \times 13 \times 9 \times \cos 10° = 19\cdot55...$
so $b = \sqrt{19\cdot55...} = 4\cdot42...$

The rock is approximately 4·4 km from the lighthouse (to 1 decimal place).

A problem involving the Sine Rule

A ship leaves port and sails 55 km on a bearing 060° to position A. Here its engines fail and it drifts to position B, which is on a bearing of 115° from the port. If position B lies on a bearing of 215° from position A, how far is the ship (at B) from the port?

Solution

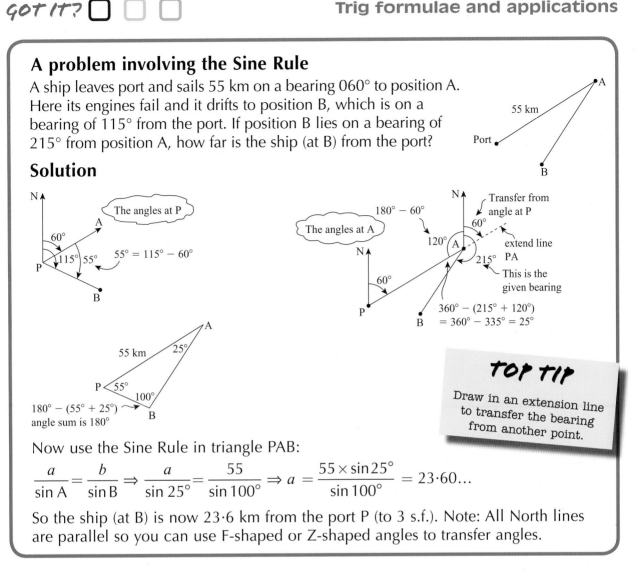

Now use the Sine Rule in triangle PAB:

$$\frac{a}{\sin A} = \frac{b}{\sin B} \Rightarrow \frac{a}{\sin 25°} = \frac{55}{\sin 100°} \Rightarrow a = \frac{55 \times \sin 25°}{\sin 100°} = 23{\cdot}60...$$

So the ship (at B) is now 23·6 km from the port P (to 3 s.f.). Note: All North lines are parallel so you can use F-shaped or Z-shaped angles to transfer angles.

TOP TIP

Draw in an extension line to transfer the bearing from another point.

Quick Test 37

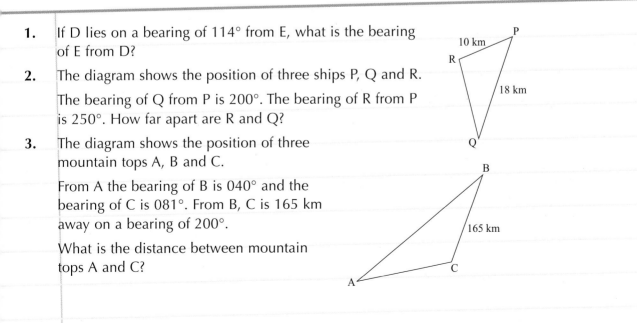

1. If D lies on a bearing of 114° from E, what is the bearing of E from D?

2. The diagram shows the position of three ships P, Q and R.

   The bearing of Q from P is 200°. The bearing of R from P is 250°. How far apart are R and Q?

3. The diagram shows the position of three mountain tops A, B and C.

   From A the bearing of B is 040° and the bearing of C is 081°. From B, C is 165 km away on a bearing of 200°.

   What is the distance between mountain tops A and C?

Working with 2D vectors

What is a vector?

A vector is a quantity with both magnitude and direction. It can be represented by a directed line segment.

Examples in real life that can be described using vectors are: momentum, velocity, push/pull forces, electric currents or fields, etc.

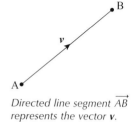

*Directed line segment $\overrightarrow{AB}$ represents the vector **v**.*

Components

TOP TIP

Vector names like **v** are printed in bold. When you write them put a line under $\underline{v}$.

Vectors are described using components parallel to the x-axis and y-axis. In the diagram, vector **v**, represented by $\overrightarrow{AB}$ has components $\binom{2}{3}$.

$$\boldsymbol{v} = \binom{2}{3} \begin{array}{l} \leftarrow x\text{-component} \\ \leftarrow y\text{-component} \end{array}$$

Think of $\binom{2}{3}$ as the 'instructions for a journey'. It describes any

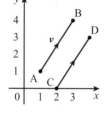

journey that goes the same distance and direction as the journey from A to B. $\overrightarrow{CD}$ also represents the same vector **v**.

Example

Write down the components of $\overrightarrow{RP}, \overrightarrow{RQ}, \overrightarrow{OS}$ and $\overrightarrow{RS}$.

Solution

$$\overrightarrow{RP} = \binom{3}{0} \quad \overrightarrow{RQ} = \binom{4}{-4} \quad \overrightarrow{OS} = \binom{-2}{-1} \quad \overrightarrow{RS} = \binom{0}{-3}$$

Note: The coordinates of S are $(-2, -1)$ and the components of $\overrightarrow{OS}$ are $\binom{-2}{-1}$.

What is a negative vector?

TOP TIP

$$\boldsymbol{v} - \boldsymbol{v} = \boldsymbol{0} = \binom{0}{0}$$

this is called the zero VECTOR. **0** is not the same as 0.

If $\overrightarrow{AB}$ represents vector **v** then...

$\overrightarrow{BA}$ represents vector $-\boldsymbol{v}$

For example, if $\boldsymbol{v} = \binom{3}{2}$ then $-\boldsymbol{v} = \binom{-3}{-2}$.

Vector addition

You can add vectors **v** and **w**

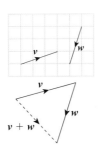

by placing them 'nose-to-tail'

or by adding their components:

$$v = \begin{pmatrix} 3 \\ 1 \end{pmatrix} \text{ and } w = \begin{pmatrix} -1 \\ -3 \end{pmatrix}$$

so $v + w = \begin{pmatrix} 3 \\ 1 \end{pmatrix} + \begin{pmatrix} -1 \\ -2 \end{pmatrix} = \begin{pmatrix} 3 + (-1) \\ 1 + (-3) \end{pmatrix} = \begin{pmatrix} 2 \\ -2 \end{pmatrix}$.

Note: $v + v$ can be written $2v = 2\begin{pmatrix} 3 \\ 1 \end{pmatrix} = \begin{pmatrix} 6 \\ 2 \end{pmatrix}$

so $2v + 3w = 2\begin{pmatrix} 3 \\ 1 \end{pmatrix} + 3\begin{pmatrix} -1 \\ -3 \end{pmatrix} = \begin{pmatrix} 6 \\ 2 \end{pmatrix} + \begin{pmatrix} -3 \\ -9 \end{pmatrix} = \begin{pmatrix} 3 \\ -7 \end{pmatrix}$.

Example

In square PQRS $\overrightarrow{PR}$ represents **a** and $\overrightarrow{RS}$ represents **b**.

Express $\overrightarrow{QR}$ in terms of **a** and **b**.

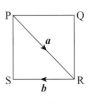

Solution

$\overrightarrow{QR} = \overrightarrow{PS}$ (they represent the same vector) and $\overrightarrow{PS} = \overrightarrow{PR} + \overrightarrow{RS}$ which represents **a** + **b** so $\overrightarrow{QR}$ represents **a** + **b**.

Vector subtraction

You can subtract vectors **v** and **w**

by placing **v** and $-w$ nose-to-tail

or by subtracting their components: $v = \begin{pmatrix} -2 \\ 2 \end{pmatrix}$, $w = \begin{pmatrix} 1 \\ 3 \end{pmatrix}$

so $v - w = \begin{pmatrix} -2 \\ 2 \end{pmatrix} - \begin{pmatrix} 1 \\ 3 \end{pmatrix} = \begin{pmatrix} -2 - 1 \\ 2 - 3 \end{pmatrix} = \begin{pmatrix} -3 \\ -1 \end{pmatrix}$.

Example:

ABCDEF is a regular hexagon with centre M.

Express:

(a) $\overrightarrow{AF}$ (b) $\overrightarrow{AE}$

in terms of **v** and **w**.

Solution

(a) $\overrightarrow{AF} = \overrightarrow{AM} + \overrightarrow{MF}$

So $\overrightarrow{AF}$ represents $-v + w$ (or $w - v$)

(b)   so

$\overrightarrow{AE} = \overrightarrow{AF} + \overrightarrow{FE}$ represents $-v + w - v = w - 2v$

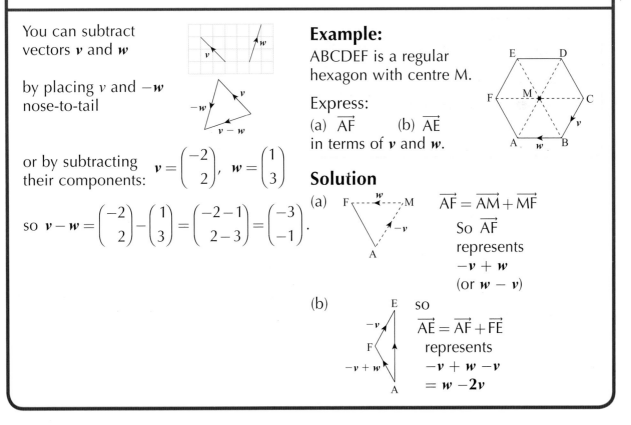

Magnitude

The magnitude (length) of a vector $\boldsymbol{v} = \begin{pmatrix} a \\ b \end{pmatrix}$ is given by: $|\boldsymbol{v}| = \sqrt{a^2 + b^2}$

(use Pythagoras' Theorem)

Example

Find the magnitude of $2\boldsymbol{a} + \boldsymbol{b}$ where $\boldsymbol{a} = \begin{pmatrix} -2 \\ 3 \end{pmatrix}$ and $\boldsymbol{b} = \begin{pmatrix} 1 \\ -2 \end{pmatrix}$.

Solution

$$2\boldsymbol{a} + \boldsymbol{b} = 2\begin{pmatrix} -2 \\ 3 \end{pmatrix} + \begin{pmatrix} 1 \\ -2 \end{pmatrix} = \begin{pmatrix} 2 \times (-2 + 1) \\ 2 \times 3 + (-2) \end{pmatrix} = \begin{pmatrix} -3 \\ 4 \end{pmatrix}$$

so $|2\boldsymbol{a} + \boldsymbol{b}| = \left| \begin{pmatrix} -3 \\ 4 \end{pmatrix} \right| = \sqrt{(-3)^2 + 4^2} = \sqrt{9 + 16} = \sqrt{25} = 5$

TOP TIP

Two short vertical lines mean 'magnitude' $|\boldsymbol{v}|$.

Quick Test 38

1. 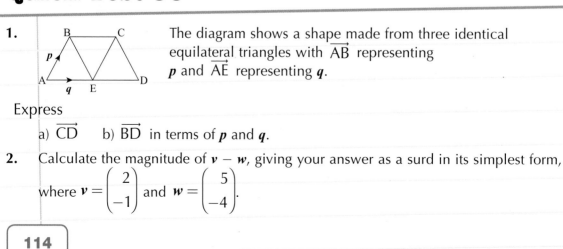 The diagram shows a shape made from three identical equilateral triangles with $\overrightarrow{AB}$ representing $\boldsymbol{p}$ and $\overrightarrow{AE}$ representing $\boldsymbol{q}$.

   Express

   a) $\overrightarrow{CD}$ b) $\overrightarrow{BD}$ in terms of $\boldsymbol{p}$ and $\boldsymbol{q}$.

2. Calculate the magnitude of $\boldsymbol{v} - \boldsymbol{w}$, giving your answer as a surd in its simplest form, where $\boldsymbol{v} = \begin{pmatrix} 2 \\ -1 \end{pmatrix}$ and $\boldsymbol{w} = \begin{pmatrix} 5 \\ -4 \end{pmatrix}$.

Coordinates and vectors in 3D

What are 3-dimensional coordinates?

To locate a point in space you need a z-axis which is at right-angles to both the x-axis and y-axis. You need three coordinates, for example the point P (1, 2, 3):

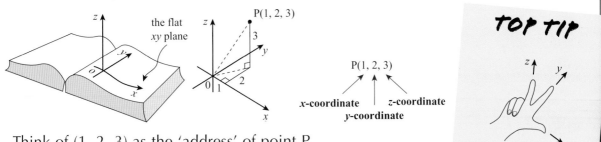

Think of (1, 2, 3) as the 'address' of point P.

TOP TIP

This is a right-hand set of axes.

Working with 3D coordinates

This diagram shows a structure built from cubes of side 1 unit. The point A has coordinates (2, 3, 4).

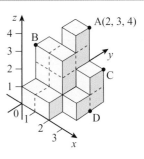

To find the coordinates of B start at the origin O.

You don't need to travel along the x-axis so the x-coordinate is 0.

You need to travel 1 unit along the y-axis, followed by 3 units up parallel to the z-axis

so B has coordinates (0, 1, 3).

In a similar way you get:

C (3, 3, 2) and D (3, 2, 0).

Example

The diagram shows a cuboid with three of its edges along the x-, y- and z-axes.

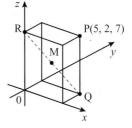

The vertex P farthest from the origin has coordinates (5, 2, 7). M is the mid point of space diagonal RQ.

Find the coordinates of M, R and Q.

Solution

R (0, 0, 7) and Q (5, 2, 0).

M is the point $\left(\frac{5}{2}, 1, \frac{7}{2}\right)$ (half of each dimension of the cuboid).

What is a 3-dimensional vector?

3D vectors are described using components parallel to the x-, y- and z-axes. In the diagram, vector $\mathbf{v}$, represented by $\overrightarrow{AB}$, has components

$$\begin{pmatrix} 1 \\ 2 \\ 3 \end{pmatrix}.$$

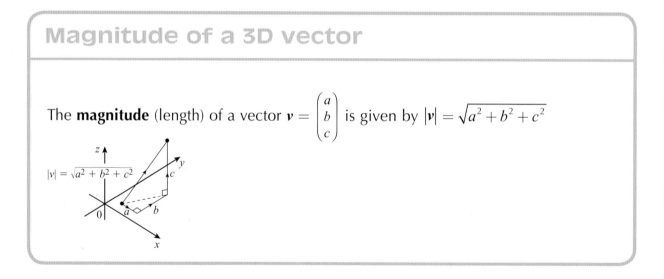

$$\mathbf{v} = \begin{pmatrix} 1 \\ 2 \\ 3 \end{pmatrix} \begin{matrix} \leftarrow x\text{-component} \\ \leftarrow y\text{-component} \\ \leftarrow z\text{-component} \end{matrix}$$

3D vectors can be added and subtracted by adding and subtracting their corresponding components. For example if

$$\mathbf{a} = \begin{pmatrix} -1 \\ 1 \\ 3 \end{pmatrix} \text{and } \mathbf{b} = \begin{pmatrix} -1 \\ -2 \\ 5 \end{pmatrix}$$

then

$$\mathbf{a} + \mathbf{b} = \begin{pmatrix} -1 \\ 1 \\ 3 \end{pmatrix} + \begin{pmatrix} -1 \\ -2 \\ 5 \end{pmatrix} = \begin{pmatrix} -1+(-1) \\ 1+(-2) \\ 3+5 \end{pmatrix} = \begin{pmatrix} -2 \\ -1 \\ 8 \end{pmatrix}$$

$$\mathbf{a} - \mathbf{b} = \begin{pmatrix} -1 \\ 1 \\ 3 \end{pmatrix} - \begin{pmatrix} -1 \\ -2 \\ 5 \end{pmatrix} = \begin{pmatrix} -1-(-1) \\ 1-(-2) \\ 3-5 \end{pmatrix} = \begin{pmatrix} 0 \\ 3 \\ -2 \end{pmatrix}$$

Magnitude of a 3D vector

The **magnitude** (length) of a vector $\mathbf{v} = \begin{pmatrix} a \\ b \\ c \end{pmatrix}$ is given by $|\mathbf{v}| = \sqrt{a^2 + b^2 + c^2}$

$|v| = \sqrt{a^2 + b^2 + c^2}$

Example

Two forces are represented by vectors

$$u = \begin{pmatrix} -3 \\ -2 \\ 0 \end{pmatrix} \text{ and } v = \begin{pmatrix} 2 \\ -3 \\ 1 \end{pmatrix}.$$

Calculate the magnitude of $u-v$.

Solution

$$u-v = \begin{pmatrix} -3 \\ -2 \\ 0 \end{pmatrix} - \begin{pmatrix} 2 \\ -3 \\ 1 \end{pmatrix} = \begin{pmatrix} -5 \\ 1 \\ -1 \end{pmatrix}$$

So $|u-v| = \left| \begin{pmatrix} -5 \\ 1 \\ -1 \end{pmatrix} \right|$

$$= \sqrt{5^2 + 1^2 + (-1)^2} = \sqrt{27}$$

$$= \sqrt{9 \times 3} = 3\sqrt{3} \text{ units}$$

(a surd in its simplest form)

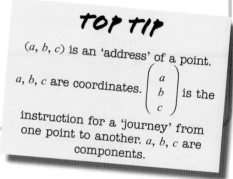

TOP TIP

(a, b, c) is an 'address' of a point.

a, b, c are coordinates. $\begin{pmatrix} a \\ b \\ c \end{pmatrix}$ is the

instruction for a 'journey' from one point to another. a, b, c are components.

Quick Test 39

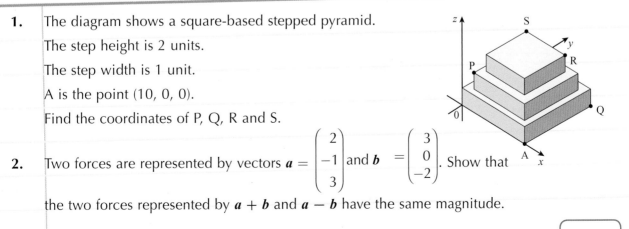

1. The diagram shows a square-based stepped pyramid.

    The step height is 2 units.

    The step width is 1 unit.

    A is the point (10, 0, 0).

    Find the coordinates of P, Q, R and S.

2. Two forces are represented by vectors $a = \begin{pmatrix} 2 \\ -1 \\ 3 \end{pmatrix}$ and $b = \begin{pmatrix} 3 \\ 0 \\ -2 \end{pmatrix}$. Show that

    the two forces represented by $a + b$ and $a - b$ have the same magnitude.

Percentages

Percentage calculations

Type 1: **Finding a % of a quantity**

Step 1 Divide the percentage number by 100.

Step 2 Multiply by the quantity.

Example

The cost of a tablet computer increased by 17·5%. Find this increase, to the nearest penny, if the cost before the increase was £525.

Solution

17·5 of £525 = $\frac{17·5}{100} \times 525$ (divided by 100, multiply by 525)
= 91·875 ≐ £91·88 (to nearest penny).

Type 2: **Expressing one quantity as a % of another quantity**

Step 1 Divide the 1st quantity by the 2nd quantity.

Step 2 Multiply by 100% (to change the fraction to a percentage).

Example

A game is sold for £14 giving a £4 profit. Express this profit as a % of the buying price.

Solution

For a £4 profit the game had a buying price of £10.

So % profit = $\frac{4}{10} \times 100\%$ (divide 4 by 10, multiply by 100%)
= 40%.

TOP TIP

Steer clear of the %
key – it is not easy to
understand!

Appreciation and depreciation

Appreciation is when a value increases. To increase £230, for example, by 15%:

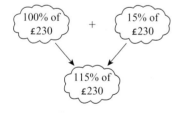

$115\% = \frac{115}{100} = 1·15$ (the multiplier)

The answer is given by:

£230 × 1·15 = £264·50

Example

A flat bought for £90 000 appreciates in value by 5% each year. What is it worth after three years? (Give your answer to the nearest £1000.)

Solution

$100\% + 5\% = 105\% = \dfrac{105}{100} = 1 \cdot 05$ (This is the multiplier.)

Value at start $= £90\,000$

After one year: $£90\,000 \times 1 \cdot 05 = £94\,500$

After two years: $£94\,500 \times 1 \cdot 05 = £99\,225$

After three years: $£99\,225 \times 1 \cdot 05 = £104\,186 \cdot 25$

Final value $=$ **£104 000** (to the nearest £1000)

Note: The calculation $90000 \times 1 \cdot 05^3$ is a quicker alternative.

TOP TIP

For appreciation, the multiplier >1. For depreciation, the multiplier <1 (but >0).

Depreciation is when a value decreases.

To decrease £230, for example, by 15%:

$85\% = \dfrac{85}{100} = 0 \cdot 85$ (the multiplier)

The answer is given by:

$£230 \times 0 \cdot 85 = £195 \cdot 50$

Example

Company shares worth £1200 depreciate in value over a month by 12% but then appreciate by 13% over the next month. Are they now worth more or less than before?

Solution

After 1st month: $£1200 \times 0 \cdot 88 = £1056$

$\quad (100\% - 12\% = 88\% = 0 \cdot 88$ is the multiplier.)

After 2nd month: $£1056 \times 1 \cdot 13 = £1193 \cdot 28$

$\quad (100\% + 13\% = 113\% = 1 \cdot 13$ is the multiplier.)

So they are worth £6·72 less.

Further percentage calculations

Type 3: Given the final amount after a % increase or decrease, finding the original amount before the change

Step 1 For an increase add the % to 100. For a decrease subtract the % from 100.

Step 2 Divide the final amount by the answer to step 1. (This calculates 1% of the amount.)

Step 3 Multiply by 100. (This calculates 100% of the original amount.)

Example
After a 15% sales reduction a TV costs £306. What was the original price?

Solution
$100\% - 15\% = 85\%$ So $85\% \leftrightarrow £306$

$1\% \leftrightarrow £\frac{306}{85}$

$100\% \leftrightarrow \frac{306}{85} \times 100 = £360$.
The original price was £360.

Compound interest

Money which is invested (called the **principal**) usually grows in value. This extra value is called **interest**.

Interest is calculated as a percentage of the principal invested. The percentage used for this calculation is called the **rate of interest**.

The letters p.a. stand for 'per annum' and mean that the interest is calculated for one complete year.

Principal + Interest = Amount

When the interest is not withdrawn but is added to the investment, it will also start to gain interest. This is called **compound interest**.

Example
Calculate the compound interest and the final amount for an investment of £950 for two years at 6% p.a.

Solution
$100\% + 6\% = 106\% = \dfrac{106}{100} = 1 \cdot 06$
(This is the multiplier.)

1st year:

amount $= £950 \times 1 \cdot 06 = £1007$

2nd year:

amount $= £1007 \times 1 \cdot 06 = £1067 \cdot 42$

So the final amount is £1067·42.

The compound interest $= £1067 \cdot 42 - £950 = £117 \cdot 42$

Note: One calculation may be used to find the amount:

$950 \times 1 \cdot 06 \times 1 \cdot 06$ or $950 \times 1 \cdot 06^2$

TOP TIP

Multiplying by, e.g. 1·06 adds on 6% of a quantity.

Quick Test 40

1. Mia's bus fare increases from £2·60 to £2·70. Calculate the % increase giving your answer to 1 decimal place.

2. A house is worth £225 000 in 2012. If it appreciates in value by 8% per year, what will its value be in 2015 to the nearest £1000?

3. Calculate (i) the final amount (ii) the compound interest when £2400 is invested for two years at 2·5% interest p.a.

4. A car is now valued at £12 600 after losing 16% of its value in a year. What was its value a year ago?

Fractions

Equivalent fractions

Identical factors in the numerator (top number) and denominator (bottom number) may be cancelled. In other words the numerator and denominator may be divided by the same number without altering the value of the fraction.

Example
Simplify: $\frac{18}{42}$.

Solution
$18 \leftarrow$ has a factor of 6

$\overline{42} \leftarrow$ has a factor of 6

$= \dfrac{\cancel{6}^{1} \times 3}{\cancel{6}_{1} \times 7} = \dfrac{3}{7}$ A shorter setting out is:

$$\dfrac{\cancel{18}^{3}}{\cancel{42}_{7}} = \dfrac{3}{7}.$$

Similarly the numerator and denominator may be multiplied by the same number without altering the value of the fraction:

$$\frac{2}{3} = \frac{6}{9} = \frac{24}{36}$$

divide top and bottom by 3   multiply top and bottom by 4

Mixed numbers

TOP TIP

If you 'read' a fraction correctly it often helps you to understand, e.g. $\frac{6}{7}$ is 6 lots of a seventh.

A number such as $3\frac{2}{5}$ is called a **mixed number**. It may be written as a 'top-heavy' fraction.

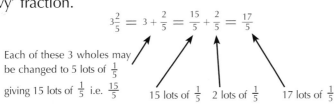

$$3\frac{2}{5} = 3 + \frac{2}{5} = \frac{15}{5} + \frac{2}{5} = \frac{17}{5}$$

Each of these 3 wholes may be changed to 5 lots of $\frac{1}{5}$ giving 15 lots of $\frac{1}{5}$ i.e. $\frac{15}{5}$

15 lots of $\frac{1}{5}$ 2 lots of $\frac{1}{5}$ 17 lots of $\frac{1}{5}$

A fraction such as $\frac{19}{3}$ which is 'top-heavy' may be written as a mixed number as follows:

$$\frac{19}{3} = \frac{18}{3} + \frac{1}{3} = 6 + \frac{1}{3} = 6\frac{1}{3}$$

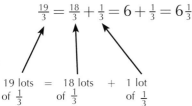

19 lots   =   18 lots   +   1 lot
of $\frac{1}{3}$ of $\frac{1}{3}$ of $\frac{1}{3}$

Note: The calculation here is 3 into 19 goes 6 times, remainder 1, so $\frac{19}{3} = 6\frac{1}{3}$.

Multiplying fractions

The basic rule is:

e.g. $\dfrac{2}{5} \times \dfrac{5}{7}$ ⟵ Multiply the two numerators.

⟵ Multiply the two denominators.

$= \dfrac{2 \times 5}{3 \times 7} = \dfrac{10}{21}$

TOP TIP

When multiplying always change mixed numbers to 'top-heavy' fractions first.

If there are identical factors in a numerator and a denominator it is wise to cancel this factor before multiplying.

Example

Evaluate: $2\frac{2}{3} \times 3\frac{3}{4}$.

Solution

$2\frac{2}{3} \times 3\frac{3}{4}$ Change these to 'top-heavy' fractions.

$= \frac{8}{3} \times \frac{15}{4}$ Now cancel the factor 3 and also the factor 4.

$= \dfrac{\cancel{8}^{2}}{\cancel{3}_{1}} \times \dfrac{\cancel{15}^{5}}{\cancel{4}_{1}} = \dfrac{2 \times 5}{1 \times 1} = \dfrac{10}{1} = 10$

Dividing fractions

After any mixed numbers are changed to 'top-heavy' fractions you should rewrite the division as a 'double-decker' fraction.

Examples of 'double-deckers' include:

$\frac{2/5}{8}, \frac{2}{1/3}$ and $\frac{23/7}{46/21}$

The numerator (top number) and the denominator (bottom number) may now be multiplied by the same number. The number you multiply by is chosen so that any cancelling reduces the 'double-decker' to a 'single-decker'. You can see this in action in the example below.

Example

Evaluate:

(a) $2 \div \frac{1}{3}$ (b) $\frac{2}{5} \div 8$ (c) $3\frac{2}{7} \div 2\frac{4}{21}$

Solution

(a) $\dfrac{2}{1/3}$

It is the 'divide by 3' that makes this a 'double-decker' so multiply top and bottom by 3 to cancel it.

$\dfrac{2 \times 3}{1/3 \times 3} = \dfrac{6}{1} = 6$

(b) $\dfrac{2/5}{8}$

This time you need to cancel the 'divide by 5' so multiply top and bottom by 5.

$= \dfrac{2/5 \times 5}{8 \times 5} = \dfrac{\cancel{2}^{1}}{\cancel{40}^{20}}$

$= \dfrac{1}{20}$

(c) $3\frac{2}{7} \div 2\frac{4}{21} = \dfrac{23}{7} \div \dfrac{46}{21} = \dfrac{23/7}{46/21}$

Multiply top and bottom by 21.

$= \dfrac{\frac{23}{\cancel{7}^{1}} \times \cancel{21}^{3}}{\frac{46}{\cancel{21}^{1}} \times \cancel{21}^{1}} = \dfrac{\cancel{23}^{1} \times 3}{\cancel{46}^{2} \times 1} = \dfrac{1 \times 3}{2 \times 1}$

$= \dfrac{3}{2} = 1\frac{1}{2}$

Adding and subtracting fractions

The aim is to get the two denominators (bottom numbers) the same.

For example: $\frac{2}{3} + \frac{4}{5}$

2 lots of $\frac{1}{3}$ 4 lots of $\frac{1}{5}$

Thirds and fifths are different so…

$$\frac{2}{3} + \frac{4}{5} = \frac{2\times5}{3\times5} + \frac{4\times3}{5\times3} = \frac{10}{15} + \frac{12}{15} = \frac{22}{15} = 1\frac{7}{15}$$

10 lots of $\frac{1}{15}$ + 12 lots of $\frac{1}{15}$ = 22 lots of $\frac{1}{15}$

Example

Evaluate: $2\frac{1}{3} + 1\frac{1}{4} - 1\frac{5}{6}$.

Solution

$2\frac{1}{3} + 1\frac{1}{4} - 1\frac{5}{6}$ Change these to 'top-heavy' fractions.

$= \frac{7}{3} + \frac{5}{4} - \frac{11}{6}$ Change $\frac{1}{3}, \frac{1}{4}$ and $\frac{1}{6}$ to twelfths.

$= \frac{7\times4}{3\times4} + \frac{5\times3}{4\times3} - \frac{11\times2}{6\times2}$

$= \frac{28}{12} + \frac{15}{12} - \frac{22}{12}$ There are $28 + 15 - 22$ lots of $\frac{1}{12}$ here.

$= \frac{28+15-22}{12} = \frac{21^{7}}{12^{4}} = \frac{7}{4} = 1\frac{3}{4}$

TOP TIP

Replace 'of' by × and always do × or ÷ calculations before + or − (beware of the order of operations).

Quick Test 41

1. Calculate (no calculator!):

    a) $1\frac{1}{2} \times \frac{4}{9}$ b) $3 \div \frac{1}{3}$ c) $1\frac{1}{2} \div 2\frac{1}{2}$ d) $\frac{6}{7} + 2\frac{1}{3}$

2. Evaluate

    a) $2\frac{1}{2} - \frac{3}{4} \times \frac{2}{3}$ b) $1\frac{1}{2} + \frac{3}{4}$ of $\frac{2}{3}$

Graphs and probability (reviews)

Types of graphs

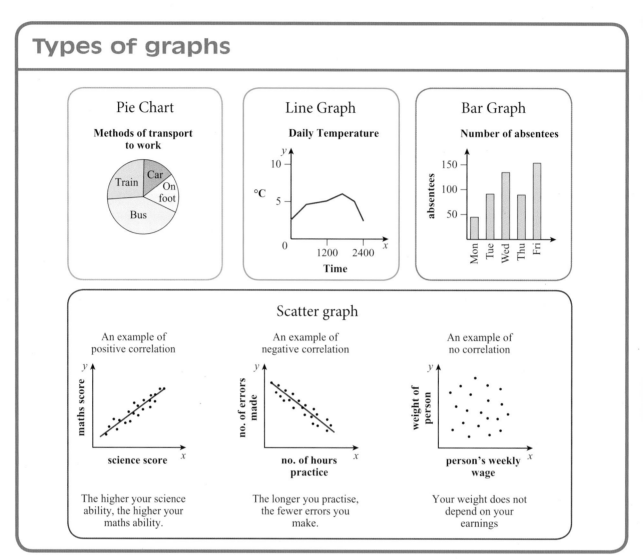

Pie Chart

Methods of transport to work

Line Graph

Daily Temperature

Bar Graph

Number of absentees

Scatter graph

An example of positive correlation

An example of negative correlation

An example of no correlation

The higher your science ability, the higher your maths ability.

The longer you practise, the fewer errors you make.

Your weight does not depend on your earnings

Pie charts and angles

In a pie chart all the data is represented by a complete circle requiring 360° at the centre. Each data value is represented by a sector of the circle with the fraction:

$$\frac{\text{angle of sector}}{360}$$

Sector

giving the fraction of data in that sector.

Example

30 students were asked what they used to write notes in class. How many used pencil?

black ink
120°
pencil
60° 180°
blue ink

Solution

$\frac{60}{360}$ ← angle for 'pencil'
    ← angle for whole circle   this is $\frac{1}{6}$

Number using pencil is $\frac{1}{6} \times 30 = 5$ students.

Example

30 commuters were asked the purpose of their journey. Thirteen said 'work', 12 were 'shopping', 3 said 'holiday', and 2 said 'visiting a friend'. Construct a pie chart to show this information.

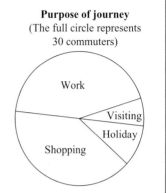

TOP TIP

A pie chart must have a clear statement of what the complete circle represents.

Solution

Here is a frequency table:

Purpose	Frequency	Fraction	Sector Angle
Work	13	$\frac{13}{30}$	$\frac{13}{30} \times 360 = 156°$
Shopping	12	$\frac{12}{30}$	$\frac{12}{30} \times 360 = 144°$
Holiday	3	$\frac{3}{30}$	$\frac{3}{30} \times 360 = 36°$
Visiting	2	$\frac{2}{30}$	$\frac{2}{30} \times 360 = 24°$

Purpose of journey
(The full circle represents
30 commuters)

Work

Visiting

Holiday

Shopping

Probability

The probability of an event happening is a number from 0 to 1:

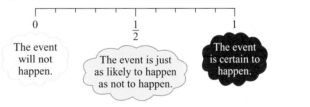

The event will not happen.

The event is just as likely to happen as not to happen.

The event is certain to happen.

In some cases your knowledge of a situation will allow you to calculate the probability of an event happening:

$$\text{Probability of an event} = \frac{\text{no. of outcomes that make the event happen}}{\text{total no. of possible outcomes in the situation}}$$

For example, if the event is: 'rolling an odd prime with a dice' then two outcomes 3 and 5 allow the event to happen. In this situation there are 6 possible outcomes (rolling 1 up to 6).

Probability of rolling an odd prime: $P(\text{odd prime}) = \frac{2}{6}$ ← favourable outcomes
                                                              ← possible outcomes

$$= \frac{1}{3}$$

If p is the probability of an event happening then $(1 - p)$ is the probability of the event **not** happening. In the example of rolling an odd prime $1 - \frac{1}{3} = \frac{2}{3}$ is the probability of not rolling an odd prime.

Probabilities can also be estimated from a sample data set, as in example 2.

Example 1

Which is more likely: drawing a face card from a pack of cards or rolling a six with a dice?

Solution

Each of the 4 suits has 3 face cards (J, Q and K) so there are 12 face cards out of 52 cards.

Probability of drawing a face card: $\frac{12}{52} \doteqdot 0{\cdot}23$ (to 2 d.p.).

Rolling a six can happen in 1 way out of 6 possible outcomes.

Probability of rolling a six $= \frac{1}{6} \doteqdot 0{\cdot}17$ (to 2 d.p.)

Drawing a face card has a higher probability and so is more likely.

Example 2

A random sample of 50 school students were asked the number of children in their families (including themselves). The results were:

No. of children in family	1	2	3	4	5	6
No. of students	8	24	16	1	0	1

Estimate the probability of a randomly picked student having only 1 brother or sister. How many such students would you expect at the school if the total roll is 1000 students?

Solution

P (family size 2) $\frac{24}{50} = 0{\cdot}48$

Expected no. at school $= 0{\cdot}48 \times 1000 = 480$

Quick Test 42

TOP TIP

Probabilities are never negative or greater than 1.

1.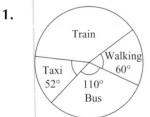

   This pie chart shows the result of a 'travel to work' survey of the 180 employees of a company. Calculate the number that take a train to work.

2. A bag contains 20 marbles, 5 each of red, blue, white and black. A marble is picked at random.

   What is the probability that it is:

   a) white   b) red or blue   c) not black   d) yellow?

3. From a bag containing numbers 1 up to 9 an even number was choosen at random.

   a) What was the probability of this happening?

   b) If the number was not replaced what is the probability of now choosing an even number from the bag?

Quartiles and averages

Quartiles

If a data set is arranged in order (smallest to largest) and written as a list on a piece of tape, the tape can be cut into four equal pieces:

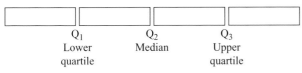

The values in the data set at the places where the tape is cut have names:

Q_1	Q_2	Q_3
Lower quartile	Median	Upper quartile

Sometimes there may be no value at the cut:

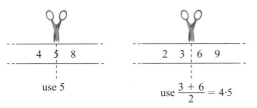

use 5   use $\frac{3+6}{2} = 4.5$

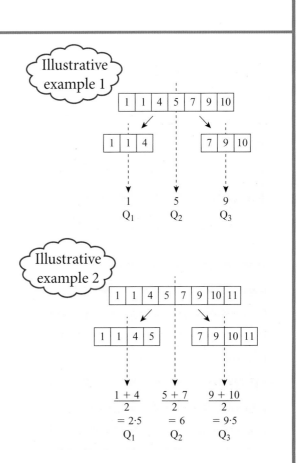

Illustrative example 1

| 1 | 1 | 4 | 5 | 7 | 9 | 10 |

| 1 | 1 | 4 | | 7 | 9 | 10 |

1   5   9
Q_1 Q_2 Q_3

Illustrative example 2

| 1 | 1 | 4 | 5 | 7 | 9 | 10 | 11 |

| 1 | 1 | 4 | 5 | | 7 | 9 | 10 | 11 |

$\frac{1+4}{2}$ $\frac{5+7}{2}$ $\frac{9+10}{2}$
$= 2.5$ $= 6$ $= 9.5$
Q_1 Q_2 Q_3

Example

A researcher counted the number of trips 17 marked bees made during the course of one day to collect nectar:

2,   3,   3,   1,   7,   2,

11,   12,   11,   6,   5,   11,

3,   7,   6,   8,   5.

Calculate the median and the upper and lower quartiles for this data set.

Solution

The data in order gives:

| 1 | 2 | 2 | 3 | 3 | 3 | 5 | 5 | 6 | 6 | 7 | 7 | 8 | 11 | 11 | 11 | 12 |

| 1 | 2 | 2 | 3 | 3 | 3 | 5 | 5 | | 6 | 7 | 7 | 8 | 11 | 11 | 11 | 12 |

$\frac{3+3}{2}$ 6 $\frac{3+11}{2}$

The median number of trips is 6, the upper quartile is 9·5 and the lower quartile is 3.

Mean and mode

TOP TIP

'Average' is a vague term. Does it mean 'mean', 'median' or 'mode'? These are all types of average.

For the data set: 2, 3, 1, 3, 1, 8.

The mean

$$\text{Mean} = \frac{2+3+1+3+1+8}{6}$$

← total of the values

← number of values

$$= \frac{18}{6} = 3$$

The mode

The mode is the most frequent value or values.

For 2, 3, 1, 3, 1, 8 there are two 1s and two 3s so there are two modes: 1 and 3.

Example

Calculate the mean and mode for the following data, which shows the number of errors made by 30 typists in a typing test:

Number of errors	0	1	2	3	4	5	6
Frequency	10	7	4	2	4	1	2

Solution

The mode is 0 (the most frequent with 10 typists).

$$\text{Mean} = \frac{10\times0+7\times1+4\times2+2\times3+4\times4+1\times5+2\times6}{30}$$

(total no. of errors)

(total number of typists)

$$= \frac{0+7+8+6+16+5+12}{30} = \frac{54}{30} = 1\cdot8$$

So the mean number of errors is 1·8.

Interquartile range

Using the lower quartile Q_1 and the upper quartile Q_3 the interquartile range is given by:

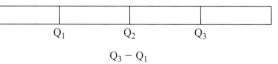

$$Q_3 - Q_1$$

It gives a measure of the distribution of data around the median Q_2. 'Clumped' data will give a relatively small number whereas 'spread out' data will give a relatively large number. It is useful when comparing two sets of data.

Example

10 girls and 10 boys were chosen at random in a school and given a test consisting of 30 subtractions. The number of errors made was recorded:

Data:

Girls: 6, 3, 0, 5, 5, 4, 0, 4, 5, 4

Boys: 9, 3, 2, 10, 6, 5, 2, 1, 7, 2

Calculate the median and the interquartile range for each of these data sets and comment on the results.

Solution

The ordered data sets are:

Girls: 0 0 3 4 4 ┊ 4 5 5 5 6

$$\frac{4+4}{2} = 4$$

3   4   5

Q_1 Q_2 Q_3

Boys: 1 2 2 2 3 ┊ 5 6 7 9 10

$$\frac{3+5}{2} = 4$$

2   4   7

Q_1 Q_2 Q_3

Interquartile ranges are:

Girls $= Q_3 - Q_1 = 5 - 3 = 2$

Boys $= Q_3 - Q_1 = 7 - 2 = 5$

Although the median number of errors is the same, 4 for both boys and girls, there is much more variation in the number of errors made by boys as indicated by the larger interquartile range for the boys (5) compared to that for the girls (2).

Quick Test 43

1. Calculate

    a) The mean and   b) the mode for the data set:

    5, 3, 8, 4, 7, 2, 3, 8, 9, 3

2. Seedling growth was tested for two new types of fertiliser, Type A and Type B. A random sample of 7 seedlings were used for Type A and 8 seedlings for Type B.

    Growth in mm:

    Type A: 8, 6, 2, 14, 2, 5, 12

    Type B: 7, 4, 3, 8, 3, 5, 8, 7

    Calculate the median and the interquartile range for each of these data sets and comment on the results.

TOP TIP

Remember you cannot calculate the median/ quartiles unless you first arrange the data in order – smallest to largest.

Standard deviation

Standard deviation–formula 1

Standard deviation is a measure of the distribution of a data set. It gives a measure of how 'spread out' the values are around their mean value.

The formula we will use here is:

$$s = \sqrt{\frac{\sum(x - \bar{x})^2}{n-1}}$$

This calculates the standard deviation, s, for a sample of values taken from a larger population. Here are the steps required to use this formula:

Step 1 Calculate the mean, $\bar{x}$, of the values:

$$\bar{x} = \frac{\text{sum of values}}{\text{number of values}} = \frac{\sum x}{n}.$$

Step 2 Calculate the deviation of each value from the mean:
value – mean = $x - \bar{x}$.

Step 3 Square each deviation:
(value – mean)^2 = $(x - \bar{x})^2$.

Step 4 Calculate the sum of these squared deviations:
Sum of squared deviations = $\sum(x - \bar{x})^2$.

Step 5 Divide the answer in Step 4 by one less than the number of values, then take the square root. This gives the standard deviation:

$$s = \sqrt{\frac{\text{sum of squared deviations}}{\text{number of values} - 1}}$$

$$= \sqrt{\frac{\sum(x - \bar{x})^2}{n-1}}$$

Example

A Quality Control Inspector selects a random sample of six matchboxes produced by a machine and records the number of matches in each:

$$52 \quad 46 \quad 50 \quad 51 \quad 49 \quad 52$$

Calculate the mean and standard deviation for this sample.

Solution

Step 1 mean $\bar{x}$

$$= \frac{52 + 46 + 50 + 51 + 49 + 52}{6}$$

$$= \frac{300}{6} = 50$$

Step 2 The deviations from the mean are:
2, −4, 0, 1, −1, 2

Step 3 The squared deviations are:
4, 16, 0, 1, 1, 4

Step 4 Sum of the squared deviations:
$4 + 16 + 0 + 1 + 1 + 4 = 26$

Step 5 The standard deviation is given by

$$s = \sqrt{\frac{\sum(x - \bar{x})^2}{n-1}}$$

$$= \sqrt{\frac{26}{5}} \quad \begin{matrix} \leftarrow \text{sum of squared deviations} \\ \leftarrow \text{there were 6 values} \end{matrix}$$

$$= \sqrt{5 \cdot 2} = 2 \cdot 3 \text{ (to 1 d.p.)}$$

The mean number of matches in a box is 50 and the standard deviation is 2·3 matches. (Usually around 95% of values are within 2 standard deviations of the mean. In this case most boxes the machine produces will contain between 45 and 55 matches.)

Table layout

It is easier to put all the calculations into a table. The table on the right shows how this can be done using the calculations from the example above.

In table form:

x Number of matches	$x - \bar{x}$ Deviations from mean	$(x - \bar{x})^2$ Squared deviations
52	2	4
46	−4	16
50	0	0
51	1	1
49	−1	1
52	2	4

> ### TOP TIP
> In exam questions it is expected that you will show clearly your working when calculating the standard deviation.

$$\sum x = 300 \qquad \sum (x - \bar{x})^2 = 26$$

$$\bar{x} = \frac{300}{6} = 50 \qquad \frac{\sum (x - \bar{x})^2}{n-1} = \frac{26}{5} = 5{\cdot}2$$

So s (standard deviation)

$$= \sqrt{5{\cdot}2} = 2{\cdot}3 \text{ (to 1 d.p.)}$$

Standard deviation – formula 2

> ### TOP TIP
> In statistical work common symbols are:
> n: number of data values
> x: a data value
> $\sum$: the total of
> $\bar{x}$: the mean
> s or σ: sample standard deviation.

The formula $s = \sqrt{\frac{\sum (x - \bar{x})^2}{n-1}}$ is one of two formulae you are given in the exam.

The other is $s = \sqrt{\dfrac{\sum x^2 - \dfrac{(\sum x)^2}{n}}{n-1}}$ and will give you the same result.

The calculations using this second formula for the previous example are shown in the table on the right.

In the exam either formula may be used.

x	x^2
52	2704
46	2116
50	2500
51	2601
49	2401
52	2704

so $\dfrac{\sum x^2 - \dfrac{(\sum x)^2}{n}}{n-1} = \dfrac{15026 - \dfrac{90000}{6}}{5} = \dfrac{15026 - 15000}{5} = 5{\cdot}2$

so $s = \sqrt{5{\cdot}2} = 2{\cdot}3$ (to 1 d.p.)

$$\sum x = 300 \quad \sum x^2 = 15026$$
$$(\sum x)^2 = 300^2 = 90000$$

Calculator advice

Scientific and graphic calculators have a statistics or STAT mode. In this mode data sets may be keyed in and various statistics calculated.

For example, after entering the data set:

$\boxed{n}$ gives the number of data values;

$\boxed{\sum x}$ gives total of the data values;

$\boxed{\bar{x}}$ gives the mean;

$\boxed{\sigma_{n-1}}$ gives the sample standard deviation.

However, be careful that you use the correct keys and only use this for checking the calculations that you have shown in your table layout. Always show working!

Quick Test 44

Zoe is comparing download prices, in pence, for music tracks. She visited six different download sites registered in the UK.

Prices per track are: 62, 59, 58, 63, 64, 60

1. Find the mean price for one track download.

2. Calculate the standard deviation for these prices to 1 decimal place.

3. Equivalent prices from US-registered sites have the same mean but with a standard deviation of 4·1. Make a valid comparison between UK download prices and US download prices.

Scatter graphs and lines of best fit

Correlation–positive or negative or none?

Scatter graphs show sets of data that involve two different quantities plotted as a set of points. If one quantity depends on the other to a certain degree then there is a correlation between the two quantities. This correlation will show on the scatter graph.

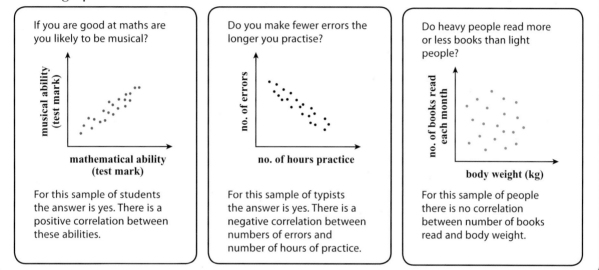

If you are good at maths are you likely to be musical?

musical ability (test mark) vs *mathematical ability (test mark)*

For this sample of students the answer is yes. There is a positive correlation between these abilities.

Do you make fewer errors the longer you practise?

no. of errors vs *no. of hours practice*

For this sample of typists the answer is yes. There is a negative correlation between numbers of errors and number of hours of practice.

Do heavy people read more or less books than light people?

no. of books read each month vs *body weight (kg)*

For this sample of people there is no correlation between number of books read and body weight.

Line of best fit

The 'line of best fit' should go approximately through the 'middle' of all the points on a scatter graph and should only be used if there is a fairly strong correlation (positive or negative).

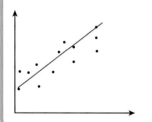 

The equation of the 'line of best fit' can be found using the methods found on pages 53 to 55. The equation can then be used to estimate values.

Remember the line is approximate and so calculations using it will only be estimates.

TOP TIP

To determine the equation of the line of best fit use two data points that are:
1. far apart
2. very close to the line.

A worked example

A sample of six fathers and their grown-up sons were weighed. The results are given in the table.

Fathers' weight (kg)	62	67	70	71	68	64
Sons' weight (kg)	63	66	67	68	66	63

(a) Draw a scatter graph for this data and describe the type of correlation involved.

(b) Draw a 'line of best fit', find its equation and use this to predict the weight of a man whose father weighed 75 kg.

Solution

(a)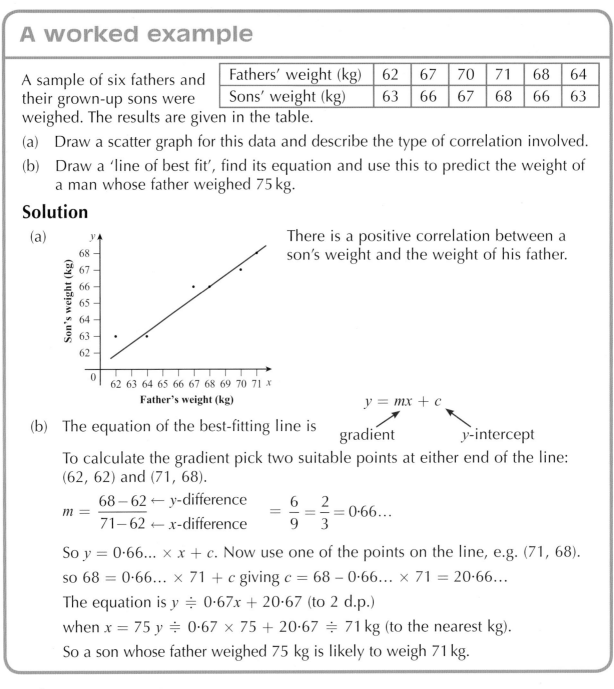

There is a positive correlation between a son's weight and the weight of his father.

(b) The equation of the best-fitting line is $y = mx + c$

gradient — y-intercept

To calculate the gradient pick two suitable points at either end of the line: (62, 62) and (71, 68).

$$m = \frac{68-62 \leftarrow y\text{-difference}}{71-62 \leftarrow x\text{-difference}} = \frac{6}{9} = \frac{2}{3} = 0.66\ldots$$

So $y = 0.66\ldots \times x + c$. Now use one of the points on the line, e.g. (71, 68).

so $68 = 0.66\ldots \times 71 + c$ giving $c = 68 - 0.66\ldots \times 71 = 20.66\ldots$

The equation is $y \doteqdot 0.67x + 20.67$ (to 2 d.p.)

when $x = 75$ $y \doteqdot 0.67 \times 75 + 20.67 \doteqdot 71$ kg (to the nearest kg).

So a son whose father weighed 75 kg is likely to weigh 71 kg.

Quick Test 45

The diagram shows a scatter graph comparing Physics score, $p\%$, with Maths score, $m\%$, gained by one particular class of students at a school. Student A scored 0% in Maths and 20% in Physics. Student B scored 90% in Maths and 80% in Physics.

1. Describe the correlation between the Physics scores and Maths scores of the students.

2. Find the equation of line of best fit AB in terms of p and m.

3. Iain was absent for the Maths test but scored 56% in the Physics test. Use the equation to estimate his Maths score.

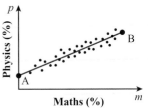

Sample unit 3 test questions

Trig problems

1. The diagram shows the measurements of a triangular field.

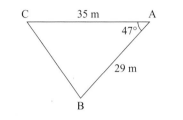

   a) Calculate the area of this field to the nearest square metre.

   b) Calculate the length of side BC (to three significant figures).

2. A coastguard station C lies on the coast 10 km due west of rock A. Rock B lies 2 km from A with angle ABC = 80°, as shown.

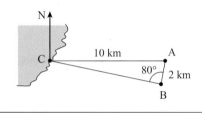

   Calculate the bearing of rock B from the coastguard station C.

Vectors

1.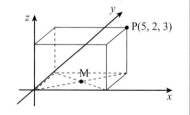

   a) Draw a directed line segment showing $a + 2b$

   b) Calculate $|a + 2b|$

2. A cuboid is placed with three of its edges along the axes as shown. P has coordinates (5, 2, 3). Write down the coordinates of M, the intersection of the two diagonals of the rectangular base of the cuboid.

3. Three forces are represented by $a = \begin{pmatrix} 2 \\ 1 \\ -1 \end{pmatrix}$, $b = \begin{pmatrix} -1 \\ 0 \\ 1 \end{pmatrix}$

   and $c = \begin{pmatrix} -3 \\ -1 \\ 0 \end{pmatrix}$ and act on an object. Find the resultant force on the object.

Fractions and percentages

1. A house valued at £230 000 is expected to increase in value by 6% each year. Calculate its expected value after three years. Give your answer to the nearest £1000.

2. An A4 sheet of paper folded into three will just fit into the envelope shown in the diagram.

   Calculate the exact area of the envelope in square inches.

   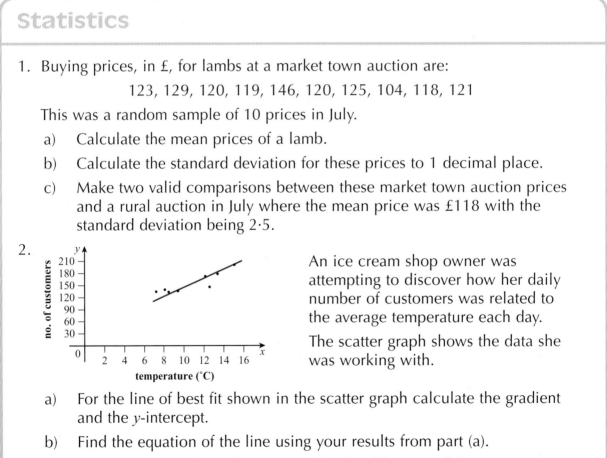

   $8\frac{5}{8}$ inches

   $4\frac{1}{4}$ inches

3. After a 15% price reduction in a sale, a tablet computer is priced £170. What was its price before the reduction?

Statistics

1. Buying prices, in £, for lambs at a market town auction are:

   123, 129, 120, 119, 146, 120, 125, 104, 118, 121

   This was a random sample of 10 prices in July.

   a) Calculate the mean prices of a lamb.

   b) Calculate the standard deviation for these prices to 1 decimal place.

   c) Make two valid comparisons between these market town auction prices and a rural auction in July where the mean price was £118 with the standard deviation being 2·5.

2. An ice cream shop owner was attempting to discover how her daily number of customers was related to the average temperature each day.

   The scatter graph shows the data she was working with.

   a) For the line of best fit shown in the scatter graph calculate the gradient and the y-intercept.

   b) Find the equation of the line using your results from part (a).

   c) Estimate the number of customers she should expect if the average temperature is 22°C.

Sample end-of-course exam questions on unit 3 topics

Non-calculator questions

1. Evaluate $\frac{3}{7}\left(1\frac{2}{3}-\frac{1}{2}\right)$.

2. Luke receives a $7\frac{1}{2}\%$ increase in salary and he now earns £8600 per year as a trainee chef. What was his salary before the increase?

3. A scatter graph is shown comparing weight, w kg, with age, m months, for a particular group of babies. Baby A when born weighed 3 kg and baby B at 16 months old weighs 5 kg.

   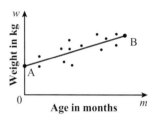

   a) Find the equation of the line of best fit AB in terms of w and m.

   b) Use the equation to estimate the age of a 3·5 kg baby from this group.

4.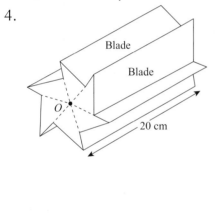

   A rotor fan in the shape of a prism has six identical blades that fit around the central point O. Each blade has a uniform cross-section with 6 cm and 4 cm sides that form a triangle meeting at O. The rotor is 20 cm in length. Given that $\sin 60° = \frac{\sqrt{3}}{2}$ calculate the exact volume of the rotor giving your answer as a surd in its simplest form.

5. Two forces represented by $\boldsymbol{p}=\begin{pmatrix} 2 \\ 4 \\ -1 \end{pmatrix}$ and $\boldsymbol{q}=\begin{pmatrix} -1 \\ 1 \\ 8 \end{pmatrix}$ are acting on a molecule.

   Calculate the magnitude of the resultant force on the molecule, writing your answer as a surd in its simplest form.

Calculator-allowed questions

1. 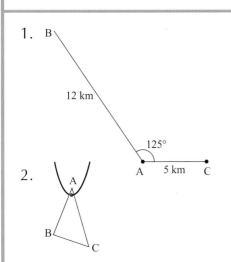 Measurements were taken from a radar display showing the position of three boats A, B and C. From A, boat B is 12 km away and boat C is 5 km away. The angle between their directions is 125°, as shown in the diagram.

   Calculate, to 3 significant figures, the distance of boat B from boat C.

2. The area of this triangular pendant is $1\frac{1}{4}$ cm². Side AB has length 2·1 cm and side AC has length 2·8 cm. Calculate the size of angle BAC to 1 decimal place.

3. A road haulage firm has a yearly running cost of £375 000. The financial department is working on a three year plan and makes the assumption that running costs will increase at the rate of 3·8% per year. Calculate, to the nearest £1000, their estimate for the running cost per year in three years time.

4. The number of hours of sunshine was recorded daily in Stonehaven over the course of a week. The results were as follows:

Sun	Mon	Tue	Wed	Thu	Fri	Sat
2·4	3·4	4·9	4·0	3·8	3·8	3·6

   a) Find the mean number of hours of sunshine that week.

   b) Find the standard deviation of the data.

   c) Kenmore, with the same mean number of hours of sunshine that week, had a standard deviation of 1·3. Make one valid comparison between the two places.

5. Three gas processing plants in the desert are connected by pipelines to form a triangle LMN, as shown. The pipelines are the following lengths:

   LM: 1·9 km LN: 3·6 km MN: 2·3 km.

   Calculate the bearing of plant N from plant M.

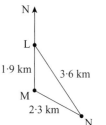

Quick Test Answers

Quick Test 1

1. a) $3\sqrt{5}$ b) $6\sqrt{6}$ c) $\dfrac{7}{8}$

2. a) $\dfrac{\sqrt{3}}{3}$ b) $2\sqrt{5}$ c) $\sqrt{2}$

3. a) $\sqrt{7}$ b) $\sqrt{5}$

Quick Test 2

1. a) 20 m   b) 0·095 km

2. a) 24   b) 3   c) 5

3. a) -3 b) 6   c) -2

Quick Test 3 (non-calculator)

1. a) 125   b) $\dfrac{1}{9}$ c) $\dfrac{4}{3}$

2. a) x^3 b) $\dfrac{1}{n^2}$ c) a^2

3. $1\cdot4 \times 10^{23}$ stars ($1\cdot44$ rounds to $1\cdot4$)

Quick Test 4

1. a) 4   b) -1

2. a) $-2x$ b) $3n^2$ c) $-3a$ d) $5b - 10a$

3. a) 81   b) 2

Quick Test 5

1. $7x - 28$ 2. $-6 + 9x$ 3. $-2x^2 - x + 1$ 4. $n^2 - n - 6$

5. $6m^2 - m - 15$ 6. $25y^2 - 20y + 4$ 7. $16 - 37a + 23a^2$

Quick Test 6

1. a) $2a(3a - 4)$ b) $7n(2m + 1)$

2. a) $(x - 7)(x + 2)$ b) $(3k - 2)(k + 3)$

3. a) $(11 - m)(11 + m)$ b) $2(3g - 7f)(3g + 7f)$

Quick Test 7

1. $(x - 6)^2 - 36$ 2. $(x + 1)^2 - 4$ 3. $\left(x - \dfrac{3}{2}\right)^2 - 2$

Quick Test 8

1. a) $5x$ b) $\frac{1}{2}$ c) $\frac{a}{a+b}$

2. a) $\frac{1}{m-n}$ b) $\frac{10}{y}$ c) $\frac{x+2}{x-2}$

Answers to quick tests

Quick Test 9

1. a) $\frac{3n}{z}$ b) $\frac{2(y-1)}{3}$ c) $\frac{m+1}{2m}$

2. a) $3x^2$ b) a c) $\frac{1}{2a^2}$

Quick Test 10

1. a) $\frac{5m-2}{m}$ b) $\frac{2-x}{x-1}$ c) $\frac{a+1}{a^2}$

2. a) $\frac{10m+7}{m(m+1)}$ b) $\frac{1}{2k}$ notice the factor $k-2$ can be cancelled!

Quick Test 11

1. a) 2   b) $-\frac{2}{3}$

2. $\frac{3}{m+n}$

3. $m_{CD} = \frac{1}{3}$, $m_{EF} = \frac{1}{3}$, $m_{CD} = m_{EF}$ so CD is parallel to EF.

Quick Test 12

1. a) 7·50 cm   b) 6·4 cm

2. a) 17·5 cm²   b) 2·1 m²

3. a) 42·6 cm   b) 61·2 m

Quick Test 13

1. a) 589 cm³   b) 12·6 cm³   c) 58·3 cm³

2. a) 3·02 cm²   b) 651 m³

Quick Test 14

1. 11   2. a) 14   b) −2   3. 7, −7

Quick Test 15

1. a) $x = \frac{1}{2}$ b) $y = \frac{3}{2}$ c) $x = -1$
2. a) $x + y = 75$ b) $2x + 3y = 195$
   where xp is the cost of an apple and yp is cost of an orange
3. a) $x < -3$ b) $x > 4$ c) $x \geq -4$

Quick Test 16

1. a) passes through $(0, -2)$ and $(4, 0)$ b) Passes through $(0, 6)$, $(2, 0)$

2. a) −2   b) $\frac{1}{3}$

3. a) Passes through $(0, 0)$ and $(3, 3)$ b) Passes through $(0, 0)$ and $(2, 3)$

Quick Test 17

1. a) $y = -\frac{1}{2}x + 2$ (or $2y + x = 4$)   b) $y = \frac{1}{3}x + 5$ (or $3y - x = 15$)

2. Line 1: A and C, Line 2: B and C

3. $y = 4x + 13$

Quick Test 18

1. $y = \frac{1}{2}x - 4$ (or $2y = x - 8$)
2. a) $t = \frac{1}{3}r + 2$ (or $3t = r + 6$)   b) $W = -2A + 7$
3. a) $2C = 5t + 6$ b) £10·50

Quick Test 19

1. a) $3x + 4y = 54$ (£x adult, £y child)

   b) $25x + 65y = 1800$ (x km bus, y km taxi)

2. a) $x + y = 10$; $2x + 3y = 24$ (£x adult, £y child)

   b) $x = 6, y = 4$

   c) Total cost is £26

Quick Test 20

1. a) $x = 2, y = 3$ b) $x = 5, y = 1$ c) $m = 6, n = -5$
2. a) $2x + 3y = 165$ b) $5x + 4y = 290$ c) $x = 30, y = 35$ so £1·60

Quick Test 21

1. a) $s = \dfrac{A}{M}$ b) $y = Pc - x$ c) $a = \pm\sqrt{c^2 - b^2}$

2. $x = \dfrac{y - b + ma}{m}$

3. $C = \dfrac{5F - 150}{9}$ or $\dfrac{5(F - 32)}{9}$

Quick Test 22

1. a) $-5, 3$ b) $\frac{1}{3}, \frac{2}{5}$
2. a) $0, 4$ b) $0, 6$
3. a) $-0·3, -3·7$ b) $0·3, 2·7$

Quick Test 23

1. $-45 = k \times 3^2$ 2. $x^2 + 10x + 28 = (x + 5)^2 + 3$
   so $k = -5$ So min turning point is $(-5, 3)$
3. $y = (3x + 2)(x - 4)$ so $x = -\frac{2}{3}$ and $x = 4$ $x = 0$ gives $y = -8$, concave upwards
   through $\left(-\frac{2}{3}, 0\right), (4, 0)$ and $(0, -8)$

Quick Test 24

1. $x = \frac{1}{2}$, max is $12\frac{1}{4}$
2. $k > -1$ (Discr $= 4 + 4k > 0$)
3. a) $(2, 1)$ b) $x = 2$.

Answers to quick tests

Quick Test 25
1. a) $3 \cdot 0$ cm b) $8 \cdot 2$ cm
2. Yes $(6^2 + 14 \cdot 4^2 = 15 \cdot 6^2 = 243 \cdot 36)$
3. 13 units
4. $5 \cdot 20$ cm (to 3 s.f.)

Quick Test 26
1. a) $a = 80, b = 45, c = 35, d = 145$ b) $e = 120, f = 60$
2. $127°$ $(53° + 74°)$

Quick Test 27
1. a) $13 \cdot 4$ cm b) $9 \cdot 17$ cm
2. 2 metres 95 cm (to the nearest cm)

Quick Test 28
1. a) $4 \cdot 5$ m b) $1 \cdot 6$ cm
2. 0.8ℓ
3. No. ($\frac{1}{8}$ th volume $\Rightarrow$ £$0 \cdot 30$ is more reasonable !)

Quick Test 29
1. a) $2 \cdot 94$ b) $20 \cdot 9$ c) $9 \cdot 59$
2. AC $= 11 \cdot 406...$ $\angle$CAD $= 46 \cdot 0°$

Quick Test 30
1. Graphs as on page 92.
2. a) $x = 270$ b) $x = 0, 180, 360$ c) $x = 180$

Quick Test 31
1. a) $k = 4, b = 2$ b) A(45, 3), B(135, −3), C(180, 0)
2. Graph $y = \cos x°$ shifted left $45°$

Quick Test 32
1. a) $41 \cdot 8, 138 \cdot 2$ b) $101 \cdot 5, 258 \cdot 5$
2. P($203 \cdot 6$, $-0 \cdot 4$), Q($336 \cdot 4$, $-0 \cdot 4$)

Quick Test 33

1. a) $t = 13$ $D = 18\cdot55$ m
    b) max: $22\cdot5$ m at 3 p.m. min: $6\cdot7$ m at 9 p.m.
2. a) $\sin^2 A + \cos^2 A - 2\sin^2 A$ (missing step)

    b) $\dfrac{\sin^2 \theta°}{\cos^2 \theta°}$ (missing step)

Quick Test 34

1. a) 136 cm²   b) 1540 cm² (use 82°)
2. $41\cdot8°$ or $138\cdot2°$

Quick Test 35

1. a) $11\cdot7$ m   b) $26\cdot3°$
2. $4\cdot53$ cm (use $\angle A = 115°$)

Quick Test 36

1. a) $8\cdot5$ m   b) $30\cdot8°$
2. $149\cdot8°$

Quick Test 37

1. $294°$ 2. $13\cdot9$ km   3. $86\cdot0$ km (to 3 s.f.)

Quick Test 38

1. a) $q - p$ (or $-p + q$)   b) $2q - p$ (or $-p + 2q$)
2. $3\sqrt{2}$

Quick Test 39

1. P(1, 1, 4), Q(15, 15, 0) R(13, 13, 6), S(2, 13, 6)

2. $|a + b| = \left\|\begin{pmatrix} 5 \\ -1 \\ 1 \end{pmatrix}\right\| = \sqrt{27}$ $|a - b| = \left\|\begin{pmatrix} -1 \\ -1 \\ 5 \end{pmatrix}\right\| = \sqrt{27}$ so $|a + b| = |a - b|$

Quick Test 40

1. $3\cdot8\%$ 2. £283 000
3. i) £2521·50   ii) £121·50
4. £15 000

Quick Test 41

1. a) $\frac{2}{3}$ b) 9   c) $\frac{3}{5}$ d) $3\frac{4}{21}$

2. a) 2   b) 2

Answers to quick tests

Quick Test 42

1. 69 travel by train
2. a) $\frac{1}{4}$ b) $\frac{1}{2}$
   c) $\frac{3}{4}$ d) 0
3. a) $\frac{4}{9}$ b) $\frac{3}{8}$

Quick Test 43

1. a) 5·2   b) 3
2. Type A: 6, 10, Type B: 6, 4 Same median, Type A more spread (*IQR* of 10 compared to *IQR* of 4)

Quick Test 44

1. 61   2. 2·4
3. More variation in US prices

Quick Test 45

1. Strong positive correlation   2. $p = \frac{2}{3}m + 20$
3. 54%